Probability and Mathematical Statistics (Continued)

RAO • Linear Statistical Inference and Its Ap
ROBERTSON, WRIGHT, and DYKSTRA •
ROGERS and WILLIAMS • Diffusions, Mark
II: Îto Calculus
ROHATGI • Statistical Inference
ROSS • Stochastic Processes
RUBINSTEIN • Simulation and the Monte Ca
RUZSA and SZEKELY • Algebraic Probability Theory
SCHEFFE • The Analysis of Variance
SEBER • Linear Regression Analysis
SEBER • Multivariate Observations
SEBER and WILD • Nonlinear Regression
SEN • Sequential Nonparametrics: Invariance Principles and Statistical Inference
SERFLING • Approximation Theorems of Mathematical Statistics
SHORACK and WELLNER • Empirical Processes with Applications to Statistics
STAUDTE and SHEATHER • Robust Estimation and Testing
STOYANOV • Counterexamples in Probability
STYAN • The Collected Papers of T. W. Anderson: 1943–1985
WHITTAKER • Graphical Models in Applied Multivariate Statistics
YANG • The Construction Theory of Denumerable Markov Processes

Applied Probability and Statistics

ABRAHAM and LEDOLTER • Statistical Methods for Forecasting
AGRESTI • Analysis of Ordinal Categorical Data
AGRESTI • Categorical Data Analysis
AICKIN • Linear Statistical Analysis of Discrete Data
ANDERSON and LOYNES • The Teaching of Practical Statistics
ANDERSON, AUQUIER, HAUCK, OAKES, VANDAELE, and WEISBERG • Statistical Methods for Comparative Studies
ARTHANARI and DODGE • Mathematical Programming in Statistics
ASMUSSEN • Applied Probability and Queues
*BAILEY • The Elements of Stochastic Processes with Applications to the Natural Sciences
BARNETT • Interpreting Multivariate Data
BARNETT and LEWIS • Outliers in Statistical Data, *Second Edition*
BARTHOLOMEW • Stochastic Models for Social Processes, *Third Edition*
BARTHOLOMEW and FORBES • Statistical Techniques for Manpower Planning
BATES and WATTS • Nonlinear Regression Analysis and Its Applications
BECK and ARNOLD • Parameter Estimation in Engineering and Science
BELSLEY • Conditioning Diagnostics: Collinearity and Weak Data in Regression
BELSLEY, KUH, and WELSCH • Regression Diagnostics: Identifying Influential Data and Sources of Collinearity
BHAT • Elements of Applied Stochastic Processes, *Second Edition*
BHATTACHARYA and WAYMIRE • Stochastic Processes with Applications
BIEMER, GROVES, LYBERG, MATHIOWETZ and SUDMAN • Measurement Errors in Surveys
BLOOMFIELD • Fourier Analysis of Time Series: An Introduction
BOLLEN • Structural Equations with Latent Variables
BOX • R. A. Fisher, the Life of a Scientist
BOX and DRAPER • Empirical Model-Building and Response Surfaces
BOX and DRAPER • Evolutionary Operation: A Statistical Method for Process Improvement
BOX, HUNTER, and HUNTER • Statistics for Experimenters: An Introduction to Design, Data Analysis, and Model Building
BROWN and HOLLANDER • Statistics: A Biomedical Introduction
BUCKLEW • Large Deviation Techniques in Decision, Simulation, and Estimation
BUNKE and BUNKE • Nonlinear Regression, Functional Relations and Robust Methods: Statistical Methods of Model Building
BUNKE and BUNKE • Statistical Inference in Linear Models, Volume I
CHAMBERS • Computational Methods for Data Analysis
CHATTERJEE and HADI • Sensitivity Analysis in Linear Regression
CHATTERJEE and PRICE • Regression Analysis by Example, *Second Edition*

*Now available in a lower priced paperback edition in the Wiley Classics Library.

Applied Probability and Statistics (Continued)

CHOW · Econometric Analysis by Control Methods
CLARKE and DISNEY · Probability and Random Processes: A First Course with Applications, *Second Edition*
COCHRAN · Sampling Techniques, *Third Edition*
COCHRAN and COX · Experimental Designs, *Second Edition*
CONOVER · Practical Nonparametric Statistics, *Second Edition*
CONOVER and IMAN · Introduction to Modern Business Statistics
CORNELL · Experiments with Mixtures, Designs, Models, and the Analysis of Mixture Data, *Second Edition*
COX · A Handbook of Introductory Statistical Methods
COX · Planning of Experiments
CRESSIE · Statistics for Spatial Data
DANIEL · Applications of Statistics to Industrial Experimentation
DANIEL · Biostatistics: A Foundation for Analysis in the Health Sciences, *Fourth Edition*
DANIEL and WOOD · Fitting Equations to Data: Computer Analysis of Multifactor Data, *Second Edition*
DAVID · Order Statistics, *Second Edition*
DAVISON · Multidimensional Scaling
DEGROOT, FIENBERG, and KADANE · Statistics and the Law
*DEMING · Sample Design in Business Research
DILLON and GOLDSTEIN · Multivariate Analysis: Methods and Applications
DODGE · Analysis of Experiments with Missing Data
DODGE and ROMIG · Sampling Inspection Tables, *Second Edition*
DOWDY and WEARDEN · Statistics for Research, *Second Edition*
DRAPER and SMITH · Applied Regression Analysis, *Second Edition*
DUNN · Basic Statistics: A Primer for the Biomedical Sciences, *Second Edition*
DUNN and CLARK · Applied Statistics: Analysis of Variance and Regression, *Second Edition*
ELANDT-JOHNSON and JOHNSON · Survival Models and Data Analysis
FLEISS · The Design and Analysis of Clinical Experiments
FLEISS · Statistical Methods for Rates and Proportions, *Second Edition*
FLEMING and HARRINGTON · Counting Processes and Survival Analysis
FLURY · Common Principal Components and Related Multivariate Models
FRANKEN, KÖNIG, ARNDT, and SCHMIDT · Queues and Point Processes
GALLANT · Nonlinear Statistical Models
GIBBONS, OLKIN, and SOBEL · Selecting and Ordering Populations: A New Statistical Methodology
GNANADESIKAN · Methods for Statistical Data Analysis of Multivariate Observations
GREENBERG and WEBSTER · Advanced Econometrics: A Bridge to the Literature
GROSS and HARRIS · Fundamentals of Queueing Theory, *Second Edition*
GROVES · Survey Errors and Survey Costs
GROVES, BIEMER, LYBERG, MASSEY, NICHOLLS, and WAKSBERG · Telephone Survey Methodology
GUPTA and PANCHAPAKESAN · Multiple Decision Procedures: Theory and Methodology of Selecting and Ranking Populations
GUTTMAN, WILKS, and HUNTER · Introductory Engineering Statistics, *Third Edition*
HAHN and MEEKER · Statistical Intervals: A Guide for Practitioners
HAHN and SHAPIRO · Statistical Models in Engineering
HALD · Statistical Tables and Formulas
HALD · Statistical Theory with Engineering Applications
HAND · Discrimination and Classification
HEIBERGER · Computation for the Analysis of Designed Experiments
HELLER · MACSYMA for Statisticians
HOAGLIN, MOSTELLER, and TUKEY · Exploring Data Tables, Trends and Shapes
HOAGLIN, MOSTELLER, and TUKEY · Understanding Robust and Exploratory Data Analysis
HOAGLIN, MOSTELLER, and TUKEY · Fundamentals of Exploratory Analyses of Variance
HOCHBERG and TAMHANE · Multiple Comparison Procedures
HOEL · Elementary Statistics, *Fourth Edition*

Continued on back end papers

*Now available in a lower priced paperback edition in the Wiley Classics Library.

Statistical Inference for Branching Processes

Statistical Inference for Branching Processes

PETER GUTTORP
University of Washington
Seattle, Washington

A Wiley-Interscience Publication
JOHN WILEY & SONS, INC.
New York • Chichester • Brisbane • Toronto • Singapore

Library of Congress Cataloging in Publication Data:

Guttorp, Peter.

Statistical inference for branching processes / Peter Guttorp.

p. cm.--(Wiley series in probability and mathematical statistics. Probability and mathematical statistics, ISSN 0271-6232)

"A Wiley-Interscience publication."

Includes bibliographical references (p.) and index.

ISBN 0-471-82291-4

1. Branching processes. 2. Mathematical statistics. I. Title. II. Series.

QA274.76.G88 1991

519.2'34--dc20 90-22224

CIP

Printed in the United States of America

10 9 8 7 6 5 4 3 2 1

To Eric, who continues the family name of
Malte, Nils, and Peter.

Preface

The purpose of this book is to illustrate, in as simple a case as possible, some of the difficulties that statistical theory, and in particular estimation theory, can encounter in the realm of dependent data. I do this by studying a very simple nonergodic model, the simple branching process discovered by Bienaymé and named after Galton and Watson. Some of the unusual features arising in this context are

- inconsistent maximum likelihood estimators
- nonestimability of many parameters of intrinsic interest
- nonnormal limit laws with classical type of norming
- nonequivalence of likelihood ratio, Wald, and score tests even under the null hypothesis
- conditional asymptotic inference is necessary

On the other hand, if one adopts a Bayesian outlook many of these problems become less important, but are usually replaced by computational difficulties.

The theory of branching processes originated from a demographic question: what is the probability that a family name becomes extinct? A brief description of the history (and solution) of this problem opens chapter 1. The family name extinction problem has colored the terminology of these processes. One talks about generations, offspring, etc. After a probabilistic construction of the process I discuss estimability of the process parameters. The main result here is due to Lockhart (1982), who showed that at most two moments and the lattice size of the offspring distribution can be estimated consistently on the explosion set, when only generation sizes and not individual offspring are observed. An intermediate scheme between these two provides for random samples of family sizes, which has important demographic applications.

Chapter 2 is the main part of the book, where results about estimating the offspring mean are presented. Estimators are derived from various criteria: the method of moments, conditional least squares, and maximum likelihood. There are a few results on finite sample properties of these estimators, and, provided that conditional inference is thought appropriate, asymptotic theory can be developed. An unusual martingale central limit theorem due to Scott (1978) yields the asymptotic distribution of these estimators.

Questions of efficiency are addressed using some general theory of inference for stochastic processes, and the conditional exponential family of Heyde and Feigin (1975) is exploited to get specific results. An important aspect of branching process inference theory is that in order to obtain frequentist results it is necessary to condition on nonextinction of the process. Such conditional results are appropriate if the event of nonextinction is ancillary, or at least approximately so. I explore the appropriateness of conditional inference in two ways. First I present the extreme point models of Lauritzen (1984), and indicate how a certain branching process fits into this framework. Then I discuss approximate and local ancillarity, and show that there are only very weak statistical arguments for conditioning on nonextinction, and much stronger arguments for conditioning on the total number of ancestors. The consequences for testing hypotheses about the offspring mean are explored (both conditionally and unconditionally), as is the random sampling of family sizes introduced in chapter 1, and inference conditional upon extinction.

In chapter 3, the estimation of parameters other than the mean is studied. The offspring variance is consistently estimable (given nonextinction), and Lockhart's result from chapter 1 is completed by exhibiting a consistent estimator of the lattice size (a parameter purely of theoretical interest, I am sure). More interesting parameters are the offspring distribution, the extinction probability, and the age of a process. The EM-algorithm of Dempster et al. (1977) can conveniently be used to estimate nonparametrically the offspring distribution (although the estimator, generally, is inconsistent), and thus any functional thereof. A particular such functional is the offspring variance. I give a consistency result (under very stringent conditions) for this estimator. Here a functional of an inconsistent estimator is consistent. In order to get consistent (or even vaguely reasonable) estimators of extinction probability or age, more stringent observational schemes, or parametric assumptions, are necessary.

A Bayesian approach is outlined in chapter 4. It avoids the conditioning problems of the frequency theory, but requires computational devices of a different kind. Asymptotics for posterior distributions and other approximation techniques are presented. Parametric models, in particular exponential family models with natural conjugate priors, as well as a nonparametric approach based on the Dirichlet process of Ferguson (1973) are analyzed.

Chapter 5 tries to illustrate how particular scientific problems require various extensions of the simple branching process. The corresponding statistical theory is outlined for each of four generalizations: branching process with immigration, with varying or random environment, with continuous state space, and multitype branching processes. I have chosen not to take the further step to continuous time processes, although these have many interesting statistical problems, since the main points of this work are most easily made in discrete time.

I have made considerable efforts to find illustrative applications that have been suggested in the literature to model concrete scientific phenomena. Whenever possible I have included the actual data, so that a reader can choose her or his own approach to analyzing them. Shorter data sets are in the text, while longer ones are given in the Appendix. As with all real data, it is not always the case that the theoretical development exactly matches the application. In some cases the application is included to illustrate the need to estimate a certain parameter, rather than to illustrate the particular estimator being developed in the text. However, this is not a book about applications of branching processes; it is concerned mainly with key concepts in estimation theory.

The results proven in this text are not always the best possible (although I have, whenever feasible, tried to indicate how results can be sharpened). This is intended to make the proofs easier to follow. Sometimes the details of an argument are deferred to the exercises at the end of each chapter. In addition I give some problems, usually from the literature, which I consider interesting and illuminating. There are also some applied exercises. I have taught a 10 week course based on some of this material to graduate students with a background of a year of measure-theoretic probability, and a year of theoretical statistics at the Ph.D. level. It helps to have encountered some applied probability, but it is not necessary. One side purpose of this book is to illustrate the power of martingales in estimation theory.

The discussion of maximum likelihood estimation of the lattice size of the offspring distribution and of the offspring variance has not appeared elsewhere. The consistency results have been obtained jointly with Richard Lockhart of Simon Fraser University. Much of the Bayes methods chapter is unpublished material. I have tried to put results from many workers in the field in a consistent perspective—aiming at a point of view rather than a complete survey.

David Brillinger told me to look at inference problems for branching processes while I was a graduate student. He and Jerzy Neyman taught me the importance of the interplay between scientific problems and statistical theory. Richard Lockhart has spent countless hours discussing many of these results with me. Niels Keiding has provided numerous helpful comments, suggestions and references. I would also like to acknowledge much help from past and present graduate students at the University of Washington, notably Michael Kahn, Daijin Ko, Michael Newton, Wasima Rida, and Anne Sheehy. My wife, June Morita, has been most supportive of this project, and no expression of my gratitude to her can suffice.

PETER GUTTORP

Seattle, Washington

Contents

CHAPTER 1

What Can Be Done?

The theory of branching processes started in the middle of the nineteenth century with social scientists thinking about the reasons for extinction of family lines. A mathematical explanation was found to render biological or social explanations superfluous. We outline the mathematical history of this problem, which contains several rediscoveries of the fundamental criticality theorem. After constructing the simple branching process probabilistically, and proving the criticality theorem, we look into which functionals of a sample path can be consistently estimated. We find the need for a concept of conditional consistency, and determine what functionals can possibly be consistently estimated, conditional upon nonextinction of the observed sample path. It is occasionally useful to take random samples from each generation of a process. Some statistical consequences of this observational scheme are discussed.

1. THE EXTINCTION OF FAMILY LINES

Social scientists have for a long time been interested in the problem of the extinction of family lines. Most of the attention has been paid to families of aristocracy or royalty, but starting in the 18th century the same phenomenon was observed among burgesses. For example, Malthus (1817) mentions that in the town of Berne, 379 out of 487 bourgeois families became extinct between 1583 and 1783. The explanation for the extinction of family lines was sought in biological terms, such as degeneration, or in social terms, such as effects of wars.

In 1845, at a meeting of the *Société Philomatique de Paris*, the noted mathematician Irénée-Jules Bienaymé presented a paper on the duration of family names. Instead of using biological or social explanations, he attempted to explain the phenomenon mathematically. Given the probability that a man has $0, 1, 2, \ldots$ sons, is it possible to compute the probability that his family line will become extinct? Bienaymé managed to relate the probability of extinction of a family name to the mean of the number of male children replacing the number of males of the preceding generation (the offspring mean). The reason for studying this problem appears to have been two papers presented (on August 31, 1844 and February 16, 1845) in the *Académie Royales des Sciences Morales et Politiques* by de Châteauneuf (1847) where observations related to family name extinction were discussed in detail. It is clear (cf. Heyde and Seneta, 1977) that Bienaymé in 1845 knew the theorem now known as the criticality theorem for branching processes. This theorem states that if the offspring mean is at most one, extinction is certain, but if the mean is larger than one, there is a computable probability (smaller than one) of extinction. This probability is given in Proposition 1.2. Bienaymé never published a formal proof of this theorem, and his work was unknown to most subsequent workers, until it was rediscovered by Heyde and Seneta (1972b).

Alphonse de Candolle published in 1873 a book, *Histoire des Sciences et des Savants*, which contains a review of the literature on extinction of aristocratic and bourgeois family lines, where he suggests that a mathematician may be able to compute the probability of extinction of a family line with known fertility distribution. He does not attempt such a computation, and is clearly unaware of Bienaymé's work.

The names Francis Galton and Henry W. Watson have traditionally been associated with simple branching processes. Their contribution came almost thirty years after Bienaymé's. Galton, who in 1869 had published the book *Hereditary Genius* where he explained the extinction of different social groups by biological causes, posed the extinction question as a problem in the *Educational Times* of 1873:

> Problem 4001. A large nation, of whom we will only concern ourselves with the adult males, N in number, and who each bear separate surnames, colonise a district. Their law of population is such that, in each generation, a_0 per cent of the adult males have no male chidren who reach adult life; a_1 have one such male child; a_2 have two; and so on up to a_5, who have five. Find (1) what proportion of the surnames will have become extinct after r generations; and (2) how many instances there will be of the same surname being held by m persons.

This problem was motivated by the publication of de Candolle's book. Upon just receiving one (incorrect) solution, Galton persuaded his acquaintance Watson ("clergyman, mathematician and alpinist" as Kendall (1966) describes him) to work out a solution. Watson obliged (Watson, 1873), using the theory of generating functions and functional iteration, but came to the incorrect conclusion that all family lines must die out. Galton wanted him to study the problem further, but had not much success. Their results were published as Watson and Galton (1874).

In 1898, the Swedish social scientist P. E. Fahlbeck published the first volume of his work on Swedish aristocratic families. He knew the results of Galton and Watson, but rejected their mathematical explanation in favor of "degeneration of reproductive ability". One of the pieces of evidence regarding this degeneration was that in extinct families, the ratio between male and female births was decreasing. However, Weinberg (1914) proved that this phenomenon *must* occur in families whose male line becomes extinct, without the necessity of a hypothesis of degeneration. Weinberg was aware of the work of Galton and Watson, but claimed to have found his results independently.

In the 1920s, Fisher (1922a) and Haldane (1927) attacked a problem in genetics using the same model as that contemplated by Bienaymé. They studied the problem of the dispersion and extinction of a mutated gene in a population (see the Applications in sections 2.5 and 2.6). Neither author refers to previous work, and Fisher's development of the theory (which did not lead as far as the criticality theorem) was evidently an independent effort. Fisher's (1930, 1958) text on genetics and natural selection has a chapter on this problem. There is no reference to Galton or Watson. Haldane's paper contains a rough outline of the criticality theorem, which is proved using a theorem on function iteration. He mentions no earlier work, except that of Fisher. The surname problem relates to genetics in that it is similar to a genetic trait which is linked to the Y-chromosome, at least in the social context of the Western world of the last few centuries.

The surname problem arise again independently in a problem a few years later, this time in *Matematisk Tidsskrift* of 1929. The poser was the Danish telecommunications engineer Agner Krarup Erlang:

> Problem nr. 15. When the probability that a person has n children is a_n, where $a_0 + a_1 + a_2 + \cdots = 1$, find the probability that his family line dies out.

Erlang, who only had managed a partial solution, died shortly after the publication of the problem. His interest arose from the fact that he was, on his mother's side, stemming from an old family, the Krarups. He was the last member of his line, and presumably interested in the extinction probability problem from that point of view. The Danish statisticians Niels Keiding and Martin Jacobsen are also members of the Krarup family. A solution to Erlang's problem was provided by J. F. Steffensen, who is credited with the first published general statement and proof of the criticality theorem. Independently, C. M. Christensen also solved the problem, and submitted a solution to the journal. It arrived later than Steffensen's solution, and was not published until 1976 (Albertsen and Kristensen, 1976). Neither of the authors were aware of the earlier work.

Another independent solution to the problem of the extinction of surnames was given by von Schelling (1944). His interest from studying average kinship between individuals in a population.

In 1986, the Computer Recreations column of *Scientific American* contained a description of a computer program developed by two British civil servants to simulate the development of aristocratic families. The offspring distribution used in the problem has a mean of 1.145. The authors found that in repeatedly running the program for 23 generations, nearly 3/4 of the families become extinct (the exact probability is 0.744), and the column author suggested that someone attempts to find out how long it takes for 1000 families to be reduced to 1, by leaving the program running for a long time. This time has positive probability of being infinite.

In order to apply the extinction probability theory, the history of which was outlined above, to real populations, one needs empirical data on offspring distributions. The first application was made by Lotka (1931), who, based on 1920 data on US white males, computed the probability of family name extinction to be 0.8797 (see the Application in section 3). A more recent computation is found in Keyfitz and Tyree (1967) and Keyfitz (1968, chapter 18), who compute the probability that the female line will become extinct for several countries, based on 1960–1961 data:

Table 1. Estimated Extinction Probabilities for Female Lines in Various Countries

USA	0.8209
Hungary	0.7130
Israel	0.5144
Mexico	0.4066
Japan	0.3242

Although the problem of extinction of family lines is the oldest problem that has been studied mathematically using branching process theory, there are several other applications. We will encounter them throughout the text. The essence of the model is that each individual (or item), independently of all other individuals, has a random number of offspring. This independence is the key probabilistic structure that enables development of the theory. Some standard references to branching process theory are Harris (1963), Athreya and Ney (1972), and Jagers (1975).

2. A PROBABILISTIC CONSTRUCTION

In this section we will formally describe the Bienaymé–Galton–Watson (BGW) branching process, and develop some interesting properties of the model. Let $X_{i,j}$, $= 1, 2, \ldots, j = 1, 2, \ldots$ be independent, identically distributed random variables, taking values in the nonnegative integers $\mathbb{N}$, with probability generating function (pgf)

$$P(s) = \sum_{k=0}^{\infty} s^k p_k .$$

We define the BGW branching process Z_k, $k = 0, 1, \ldots$ with offspring distribution P recursively by

$$Z_0 = z_0 , \qquad Z_k = \sum_{i=1}^{Z_{k-1}} X_{i,k} . \tag{1}$$

Unless specifically stated we will assume that $z_0 = 1$, and that $p_0 + p_1 < 1$. The conditional pgf of Z_k, given Z_{k-1} is by independence

$$\mathbf{E}(s^{Z_k} | Z_{k-1} = z) = P(s)^z ,$$

so that pgf of Z_k is

$$P_k(s) = \mathbf{E}P(s)^{Z_{k-1}} = P_{k-1}(P(s)) . \tag{2}$$

From (2) we deduce that

$$\mathbf{E}Z_k = m^k , \tag{3}$$

provided $m = \mathbf{E}X_{i,j} < \infty$, and

$$\mathbf{Var}\, Z_k = \sigma^2 m^{k-1} \sum_{0}^{k-1} m^i , \tag{4}$$

provided $\sigma^2 = \mathbf{Var}\, X_{i,j} < \infty$.

It is clear from (1) that $(Z_k)_{k \geq 0}$ is a Markov chain with transition probabilities given by

$$p(i|j) = \mathbf{Pr}\{Z_n = i | Z_{n-1} = j\} = p_i^{*j}$$

where $p_\cdot^{*j}$ are the probabilities of a j-fold convolution of the offspring distribution. A standard argument (see, e.g., Freedman, 1971, pp. 2–3) shows the existence of a probability space $(\Omega, \mathcal{F}, \mathbf{P})$ for which the coordinate process is Markov with transitions $p(i|j)$. This construction, however, does not have enough structure to assure that interesting random variables, such as "the number of fifth generation offspring of the third particle of the second generation" can be defined. The representation (1) is adequate for such purposes. It will, however, prove instructive to construct a slightly larger probability space than the one naturally associated with (1).

Let $J = \{\iota\}$ be a set of indices, where ι is either 0 or a sequence $i_1 i_2 \ldots i_k$ of positive integers. We enumerate J in some fixed order, for example

$$0121132112111431222111312111121111 \ldots .$$

We denote individuals by (ι), where (0) is the *ancestor* with children (1), (2) $\ldots (\iota) \ldots$ The symbol (ij) stands for the jth child of the ith child of the ancestor, etc. Define a *family history* by

$$\omega = (\nu_0, \nu_1, \omega_2, \nu_{11}, \nu_3, \ldots)$$

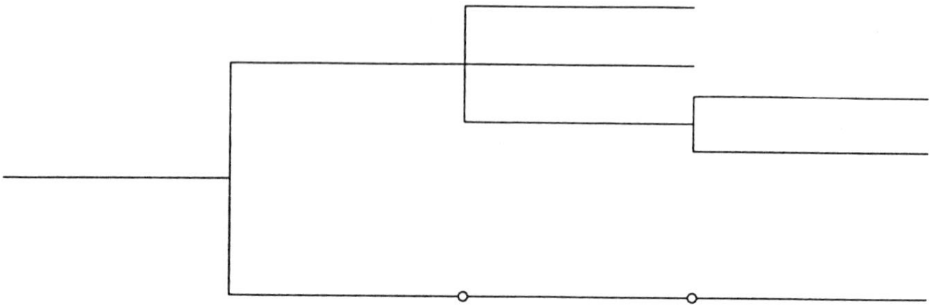

FIGURE 1.1. A family tree with $\nu_0 = 2$, $\nu_1 = 3$, $\nu_2 = 1$, $\nu_{21} = 1$, and $\nu_{13} = 2$.

where ν_ι is the number of children of (ι). In terms of the representation (1) we have that $\nu_0 = X_{1,1}$, and for $k \geq 1$, $\nu_{i_1 \ldots i_k} = X_{i,k}$ for some i. Notice that if for exampe $\omega_2 = 0$, then $\nu_{2i_1 \ldots i_k} = 0$ for all $i_1 \ldots i_k$. Define a *family tree* as the realized part of ω, i.e., the nonzero elements. This can be drawn in tree form as shown in Figure 1.1. Clearly, (ij) is in the tree if and only if $i \leq \nu_0$ and $j \leq \nu_i$. Define the kth generation recursively by

$$I_k(\omega) = \{(i_1 \ldots i_k) \colon (i_1 \ldots i_{k-1}) \in I_{k-1}(\omega) \text{ and } \omega_{i_1 \ldots i_{k-1}}(\omega) \geq i_k\}$$

and let the *family* be $I(\omega) = \bigcup_k I_k(\omega)$. Notice that I may be infinite, though each I_k will be finite. Now let $\Omega = \{\omega\}$ and define $\tilde{\mathbf{P}}$ on Ω by making the ν_ι independent and identically distributed, according to the offspring pgf $P(s)$. Define $Z_k(\omega) = \#(I_k(\omega))$. Let $\mathbf{P}$ be the distribution of $(Z_k, k \geq 0)$ induced by $\tilde{\mathbf{P}}$.

Proposition 1.1: $(Z_k)_{k \geq 0}$ is a BGW branching process on $(\Omega, \mathcal{F}, \tilde{\mathbf{P}})$ where $\mathcal{F}$ is the σ-algebra generated by the cylinder sets of Ω.

Proof: We must show that

$$\tilde{\mathbf{E}}(s^{Z_k} | Z_1, \ldots, Z_{k-1}) = P(s)^{Z_{k-1}},$$

where $\tilde{\mathbf{E}}$ is the expectation corresponding to $\tilde{\mathbf{P}}$. Let $F = \{Z_1 = z_1, \ldots, Z_{k-1} = z_{k-1}\}$, and

$$F_{\mathbf{n}} = \left\{ \begin{array}{l} \nu_0 = z_1, \\ \nu_1 = n_1, \ldots, \nu_{z_1} = n_{z_1}, \\ \nu_{11} = n_{11}, \ldots, \nu_{1n_1} = n_{1n_1}, \\ \vdots \\ \nu_{z_1 1} = n_{z_1 1}, \ldots, \nu_{z_1 n(z_1)} = n_{z_1 n(z_1)}, \\ \vdots \\ \nu_{11\ldots1} = n_{11\ldots1}, \ldots, \nu_{z_1 n(z_1(\ldots n(z_1)\ldots))} = n_{z_1 n(z_1(\ldots n(z_1)\ldots))} \end{array} \right\}$$

where the indices on the last line correspond to individuals in the generation $k-1$, and parentheses replace subscripting for readability. Notice that

$$z_k = \sum_{i_0=1}^{\nu_0} \sum_{i_1=1}^{\nu_{i_0}} \cdots \sum_{i_{k-1}=1}^{\nu_{i_0 \dots i_{k-2}}} \nu_{i_0 \dots i_{k-1}} .$$

Then

$$\tilde{\mathbf{E}}(s^{Z_k}|F) = \sum \tilde{\mathbf{E}}(s^{\sum_{i_0} \dots \sum_{i_{k-1}} \nu(i_0 \dots i_{k-1})} |F_{\mathbf{n}})\tilde{\mathbf{P}}(F_{\mathbf{n}}|F)$$

where the sum is taken over all integers $n_1, \dots, n(z_1(n(z_2) \dots n(z_{k-2})))$ such that

$$n_1 + \cdots + n_{z_1} = z_2, \dots, n_{11\dots1} + \cdots + n(z_1(n(z_2) \dots n(z_{k-2}))) = z_{k-1} .$$

But this conditional expectation is simply $P(s)^{Z_{k-1}}$, since the νs are all independent and identically distributed and there are precisely z_{k-1} of them. Also, $\Sigma\, \tilde{\mathbf{P}}(F_{\mathbf{n}}|F) = 1$, whence the result follows. □

In what follows, the measure **P**, constructed from $\tilde{\mathbf{P}}$ above, will be called the distribution of the BGW process (Z_k). **P** is a measure on the observed σ-algebra $\mathcal{F}^Z = \sigma(Z_k,\ k \geq 0)$. In other words, we will typically restrict ourselves to observing generation sizes, and not allow the observation of individual offspring.

The following result often comes in handy.

Lemma 1.1: Let $(Z_k^{(j)})$, $j = 1, \dots, N$ be independent BGW processes each starting with one ancestor and with the same offspring distribution as (Z_k). Then for any $F \in \mathcal{F}^Z$

$$\mathbf{P}[\{Z_k; k > r\} \in F | Z_r = N] = \mathbf{P}\left[\left\{\sum_{j=1}^{N} Z_k^{(j)}, k \geq 1\right\} \in F\right].$$

Proof: Let $F = \{r_1\} \times \cdots \times \{r_k\} \times \mathbb{N} \times \mathbb{N} \times \cdots$. Taking disjoint unions of such F yields all finite-dimensional sets in $\mathcal{F}^Z$, which generate $\mathcal{F}^Z$, so we just need to verify the result of this F. The essential idea can be seen for $k = 1$. Suppose the individuals in the rth generation of the original process are $\iota_1, \dots, \iota_N$. Then $Z_{r+1} = \Sigma_{j=1}^{N} \nu_{\iota_j}$. This has the same distribution as $\Sigma_{j=1}^{N} \nu_0^{(j)} = \Sigma_{j=1}^{N} Z_1^{(j)}$ by the construction in Proposition 1.1, which also can be used to get the result for general k. □

3. ASYMPTOTIC BEHAVIOR

From the discussion about the construction in the previous section it should be clear that once a generation is empty, all following generations will be empty as well. Therefore we will compute the probability of eventual extinction. Extinction will occur in or before the kth generation is one of the following ways: the ancestor has

0 children
1 child whose family become extinct in or before the $(k-1)$th generation
2 children, both of whose families become extinct in or before the $(k-1)$th generation, etc.

Since the probability of the jth of these cases is p_j, the probability q_k of extinction after k generations can be computed using Lemma 1.1 as

$$q_k = \mathbf{P}\{Z_k = 0\} = P_k(0) = \sum_{j=0}^{\infty} p_j P_{k-1}(0)^j = P(P_{k-1}(0)) = P(q_{k-1}) . \qquad (5)$$

Assuming that $p_0 \neq 0$ and that $p_1 \neq 1$, we note that $P(s)$ is a strictly increasing function. Thus the numbers $q_k = P(q_{k-1})$ form a strictly increasing sequence of positive numbers, all bounded by 1. This sequence therefore has a limit which we denote q, with $p_0 \leq q \leq 1$. This is the probability of ultimate extinction. By going to the limit in (5), we see that $q = P(q)$, whence

$$\frac{P(q) - P(q_k)}{q - q_k} = \frac{q - q_{k+1}}{q - q_k} < 1 .$$

Letting $k \to \infty$ we see that $P'(q) \leq 1$. Since $P'(s)$ is a power series with positive coefficients it is an increasing function, i.e. $P(s)$ is convex. If $P'(1) > 1$ we must have $q < 1$. In this case q and 1 are the only positive roots to the equation $P(s) = s$. On the other hand, if $P'(1) \leq 1$, we have for $0 \leq s < 1$ that $P'(s) - 1 < 0$, so that the smallest zero of the function $P(s) - s$ must be 1, whence $q = 1$ (see Figure 1.2).

Finally, if $p_0 = 0$, $Z_k \geq Z_{k-1}$, and extinction is impossible. If $p_1 = 1$, $Z_k = Z_0$, and extinction is again impossible. Upon noting that $P'(1) = m$, we have proved the following.

Proposition 1.2: The extinction probability q is the smallest nonnegative root to the equation

$$P(s) = s .$$

If $m \leq 1$ this probability is 1, unless $p_1 = 1$ when it is 0. If $m > 1$ we have that $0 \leq q < 1$, with equality if and only if $p_0 = 0$.

Remark: Watson's solution to the family name extinction problem correctly identified q as a solution to $P(s) = s$. However, he failed to realize that there could be a solution other than 1. □

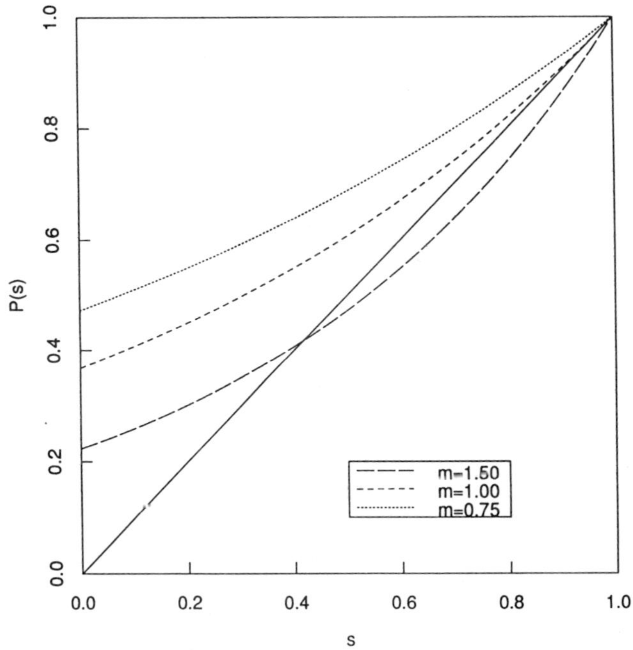

FIGURE 1.2. Three different pgfs $P(s)$, one with $P'(1) > 1$ (long dashed), one with $P'(1) = 1$ (short dashed), and one with $P'(1) < 1$ (dotted). The line $y = s$ is also shown (solid). q is the smallest point of intersection between the curve and the line.

Example: The Geometric Distribution

Throughout this book we will illustrate the application of various ideas to a particularly simple case, the *geometric* offspring distribution, with $p_k = p(1-p)^k$. We have $P(s) = p/(1-(1-p)s)$, so $m = P'(1) = (1-p)/p$. From Proposition 1.2 we see that this process becomes extinct with probability 1 whenever $p \geq 1/2$. When $p < 1/2$ the extinction probability solves the equation

$$\frac{p}{1-(1-p)s} = s$$

with solutions 1 and $1/m$. Hence $q = 1/m$.

Exercise 3 indicates how a geometric distribution with a modified zero term is maintained for a branching process, in the sense that if this is the offspring distribution, then each generation size also has a distribution of the same type (but with different parameters). This offspring distribution is not only mathematically convenient, it also arises naturally from discrete observations of a birth and death process. Let $X(t)$ be a temporally homogeneous linear birth and death process. This is a Markov process where the probability of a unit increase in $(t, t+dt)$ is $X(t)\lambda\, dt + o(dt)$, and the probability of a unit decrease in $(t, t+dt)$ is $X(t)\mu\, dt + o(dt)$. Let

$$P_{ij}(t) = \mathbf{P}(X(s+t) = j \mid X(s) = i)\,.$$

From the Kolmogorov forward equation (e.g., Cox and Miller, 1965, Example 4.4)

$$\frac{\partial P_{ik}(t)}{\partial t} = -k(\mu + \lambda)P_{ij}(t) + \lambda(k-1)P_{i,k-1}(t) + \mu(k+1)P_{i,k+1}(t)$$

one gets

$$P_{10}(t) = \frac{\mu(1 - \exp(-t(\mu - \lambda)))}{\mu - \lambda \exp(-t(\mu - \lambda))}$$

and

$$P_{1n}(t) = (1 - P_{10}(t))\eta(t)(1 - \eta(t))^{n-1}$$

where

$$\eta(t) = \frac{\lambda(1 - \exp(-t(\mu - \lambda)))}{\mu - \lambda \exp(-t(\mu - \lambda))}\,.$$

If we rewrite the forward equation in terms of the generating function $F_i(s, t) = \Sigma_{k=0}^{\infty} P_{ik}(t)s^k$ we get

$$\frac{\partial F_i(s, t)}{\partial t} = (\mu - (\lambda + \mu)s + \lambda s^2)\,\frac{\partial F_i(s, t)}{\partial s}\,. \tag{6}$$

Suppose now that $F_1(s, t)$ is a solution to this equation with initial value $F_1(s, 0) = s$. Consider the function F_1^i. It is easy to see that this function also satisfies (6), with initial value s^i. Hence $F_i(s, t) = F_1(s, t)^i$. Now consider the process $X(t)$ observed at discrete times $k\Delta$. If we write $Z_k = X(k\Delta)$ we see that

$$\begin{aligned}\mathbf{E}(s^{Z_{n+1}} \mid Z_n = i) &= \mathbf{E}(s^{X(\Delta)} \mid X(0) = i) = F_i(s, \Delta)\\ &= F_1(s, \Delta)^i = \mathbf{E}(s^{Z_1} \mid Z_0 = 1)^i\,.\end{aligned}$$

Thus (Z_k) is a BGW process, with offspring distribution $P_{1j}(\Delta)$, i.e., a geometric distribution with modified zero term. □

We now introduce the *extinction set* $B = \bigcup_k \{\omega \colon I_k(\omega) = \varnothing\}$. Clearly

$$\mathbf{P}(B) = q\,.$$

Exercise 4 states the obvious fact that if the process does not become extinct, then all generations must have at least one individual. In fact, more can be said.

Proposition 1.3: Provided that $p_1 \neq 1$, $Z_n \to \infty$ with probability $1 - q$.

Proof: If $q = 1$ there is nothing to prove, so we let $q < 1$. First note that $P_2'(q) = P'(q)^2$, and that, by induction, $P_n'(q) = P'(q)^n$. We estimate, given k, that

$$\mathbf{P}\{1 \le Z_n \le k\} \le \sum_1^k \mathbf{P}\{Z_n = j\} \frac{jq^{j-1}}{q^k}$$
$$\le \frac{P_n'(q)}{q^k} = \frac{P'(q)^n}{q^k} .$$

Summing over n, the Borel–Cantelli lemma shows that

$$\mathbf{P}\{1 \le Z_n \le k \text{ i.o.}\} = 0 ,$$

for any value of k. □

Remarks

(i) Billingsley (1962) developed the theory of statistical inference for recurrent Markov chains. The preceding proof shows that BGW processes have two classes of states: the zero state, which is absorbing, and the other states, which are all transient. Hence Billingsley's results do not apply, and we will have to develop some different notions of statistical inference.

(ii) The criticality theorem (in common usage) is really the combination of Propositions 1.2 and 1.3, and says that a BGW process either becomes extinct, with a probability depending on m, or explodes. If $m < 1$ the process is called *subcritical*, while if $m = 1$ it is *critical* and if $m > 1$ it is *supercritical*. □

Application: The Extinction of Family Names

In order to develop a better feel for the behavior of these processes, we will use Lotka's (1931) data on white males in the United States in 1920 as an illustration. The observed data are

$$p_0 = 0.4982 \qquad p_1 = 0.2103 \qquad p_2 = 0.1270$$
$$p_3 = 0.0730 \qquad p_4 = 0.0418 \qquad p_5 = 0.0241$$
$$p_6 = 0.0132 \qquad p_7 = 0.0069 \qquad p_8 = 0.0035$$
$$p_9 = 0.0015 \qquad p_{10} = 0.0005 \qquad p_n = 0.0000 \quad (n > 10)$$

The exact extinction probability, obtained by solving a polynomial equation of degree 10, is 0.8798. Looking at successive ratios of the probabilities,

Lotka found that a geometric distribution with modified zero term (see Exercise 3), was a good fit to the data. The maximum likelihood estimates of the parameters (see the second Example in section 2.4) are

$$\hat{c} = 0.5617\,, \qquad \hat{b} = 0.2199\,,$$

corresponding to an offspring mean of 1.145 and an extinction probability of 0.8869. These values differ from those used by Lotka, who fit a least squares line to the logarithm of the observed frequencies. The fit using the mles is better. The corresponding pgf is

$$P(s) = 0.4982 + \frac{0.2199s}{1 - 0.5617s}\,.$$

Using Exercise 3(d) we can compute the expected time until extinction of a unique family name, conditional upon the extinction occurring. This is seen to be 2.3 generations, corresponding to about 70 years. de Châteauneuf observed a mean survival of noble family names of 300 years, or about ten generations. The 300 year mean survival of the French nobility is perhaps not a reasonable number for white American males. Assuming (which is of course not very realistic) that the offspring distribution of white American males stays the same for three generations, we find that the extinction probability after three generations, q_3, is 0.724, so that only one family name in four (each family starting with one member in 1920) would survive to the next century. Since the average family size for white families in the United States has been decreasing (Blake, 1989), this is an overestimate.

If we study a particular family name, unique in 1920, which survives into the twenty-first century, the conditional distribution of the number of male descendants after three generations (given survival) is geometric with parameter

$$(1 - q)/(m^3 - q) = 0.1842$$

which has mean 5.4. This distribution is quite long-tailed, as seen in Figure 1.3. One in twenty surviving families would have at least 14 members, one in six would have only one. Therefore the mean is of limited descriptive use in this case.

If we look at what would happen in ten generations, we find $q_{10} = 0.853$, and the mean family size after ten generations would be $m^{10} = 3.87$. In other words, the male half of the population would be increased fourfold, but the number of different surnames would be decreased sevenfold. □

Using martingale theory we can say more about the behavior of paths of a BGW branching process.

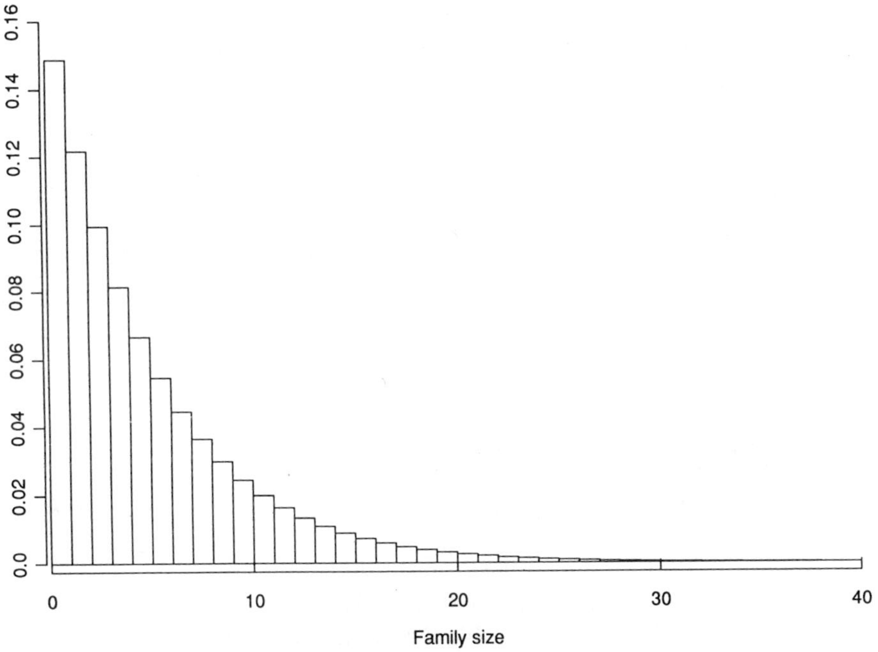

FIGURE 1.3. The conditional distribution of the number of male descendants after three generations, given survival of a family name unique in 1920 among white American males.

Theorem 1.1: Let $W_n = m^{-n} Z_n$, and let $\mathscr{F}_n = \sigma(Z_0, \ldots, Z_n)$. Assume that $\sigma^2 = \mathbf{Var}\, Z_1 < \infty$. Then

(i) W_n is an $\mathscr{F}_n$-martingale.
(ii) $W_n \to W$ a.s. where $\mathbf{P}(W = 0) = q$.
(iii) $\{W > 0\} = \{Z_n \to \infty\}$ a.s.
(iv) If $m > 1$, then $\mathbf{E}W = 1$ and $\mathbf{Var}\, W = \sigma^2 / m(m-1)$.
(v) The Laplace transform $\phi(s) = \mathbf{E}e^{-sW}$ of the distribution of W satisfies Abel's equation

$$\phi(ms) = P(\phi(s)), \qquad s \geq 0.$$

(vi) If $m > 1$ and $\sigma^2 < \infty$, the distribution of W is absolutely continuous except for a jump of size q at 0.

Proof: (i) $\mathbf{E}|W_n| = 1$, and

$$\mathbf{E}(W_{n+1} | \mathscr{F}_n) = \mathbf{E}(m^{-(n+1)} Z_{n+1} | Z_n) = m^{-n} Z_n = W_n .$$

The convergence in (ii) follows from the martingale convergence theorem, since $\sup_n \mathbf{E}|W_n| = 1$. The probability statement is a consequence of (iii).

To prove (iii), first notice that $Z_n \to 0$ implies that $W = 0$. Conversely, $\mathbf{P}(W_n \to 0 | Z_1 = k) = (\mathbf{P}(W_n \to 0))^k$ (using Exercise 1). Averaging over the distribution of Z_k we see that if $\pi = \mathbf{P}(W = 0)$, then $\pi = P(\pi)$, so $\pi = q$ or 1. We must rule out $\pi = 1$ if $m > 1$. This follows from (iv), since $\mathbf{E}W = 1$ is impossible if $\mathbf{P}(W = 0) = 1$.

For part (iv), W_n converges in L_2, since

$$\mathbf{E}(W_{n+k} - W_n)^2 = \sigma^2 m^{-n} \frac{1 - m^{-k}}{m(m-1)} .$$

Since $\mathbf{E}W_n = 1$ and $\mathbf{Var}\ W_n \to \sigma^2/m(m-1)$ the result follows.

To verify (v), let $\phi_n(s) = \mathbf{E}e^{-sW_n} = P_n(e^{-s/m^n})$. From Exercise 1 it follows that

$$\phi_{n+1}(ms) = P(\phi_n(s)) , \qquad s \geq 0, n = 1, 2, \ldots .$$

Now $W_n \to W$ in distribution, so the Laplace transforms $\phi_n \to \phi$. Since P is a continuous function, the statement follows.

We will just outline a proof of (vi) assuming that $P(0) = 0$ (see Athreya and Ney, 1972, Theorem I.10.4 and Corollary I.12.1 for a complete and general version). The equation in (v) can be iterated to show that $\phi(im^n t) = P_n(\phi(it))$. Since $q = 0$, the result of Exercise 7 implies that $\phi(it) \to 0$ as $|t| \to \infty$. A careful analysis of the derivative of the same equation shows that $\phi'(t)$ is absolutely integrable. By the Fourier inversion theorem, W has a density

$$f_W(w) = \frac{1}{2\pi w} \int_{-\infty}^{\infty} \exp(-itw)\phi'(-it)\, dt . \qquad \square$$

Remark: The assumption that $\sigma^2 < \infty$ is stronger than needed for Theorem 1.1 (except, of course, for the variance part of (iv)). Seneta (1969) and Heyde (1970) showed that for a supercritical BGW process there are constants $c_n > 0$, $c_n \to \infty$, $c_{n+1}/c_n \to m$ and such that $c_n^{-1} Z_n \to W$ a.s., where W is a proper nondegenerate random variable. Furthermore $\mathbf{E}X \log X < \infty$ if and only if $c_n = O(m^n)$, where K is a constant. Cohn (1982) shows that the c_n may be chosen as the μ-quantile of the distribution of Z_n, where $q < \mu < 1$. Athreya and Ney (1972) show that for supercritical processes $\mathbf{E}W = 1$ whenever $\mathbf{E}Z_1 \log Z_1 < \infty$. If the latter expectation is infinite, $W = 0$ with probability 1. $\square$

Example: The Positive Geometric Distribution

In the positive geometric case (where there is no possibility of extinction) it is easy to compute the distribution of W. The equation in (v) is

$$\phi(s/p) = \frac{p\phi(s)}{1 - (1-p)\phi(s)} .$$

A solution is $\phi(s) = \lambda/(\lambda + s)$, i.e., $W \sim \exp(1)$, using (iv). By Exercise 9 this solution is unique. □

Corollary: Let $Y_n = \Sigma_0^n Z_i$. If $m > 1$, $m^{-n} Y_n \to mW/(m-1)$ a.s.

Proof: The proof is based on the Toeplitz lemma.

Lemma 1.2: Let (a_k) be a sequence of real numbers, and set $b_n = \Sigma_{k=1}^n a_k$. Suppose $x_n \to x < \infty$. Then

$$\frac{1}{b_n} \sum_{k=1}^{n} a_k x_k \to x .$$

Since $m^{-n} Z_n \to W$ a.s. we get by the lemma that

$$\sum_{k=0}^{n} m^k (m^{-k} Z_k - W) \Big/ \sum_{k=0}^{n} m^k \to 0 \text{ a.s.}$$

so that

$$\left(\sum m^k\right)^{-1} \sum Z_k \to W \text{ a.s.}$$

But $\Sigma_{k=0}^n m^k = [m/(m-1)] m^n (1 + o(m^{-n}))$, whence the claim follows. □

4. NONEXISTENCE OF CONSISTENT ESTIMATORS

In classical iid statistical theory *consistency* is an important requirement of an estimator. If we have a large sample, any reasonable estimator should be close to the true parameter. In the case of stochastic processes, the large sample is often obtained by letting the observation time get large, i.e., by looking at a whole path of the process. In this section we will prove a simple theorem to the effect that unless $p_0 = 0$ we cannot find any consistent estimators based on a single infinite realization of a BGW branching process. The problem is that the process may become extinct, in which case there is not enough information in a path to tell two distinct probability measures apart. In other words, even when the observation time becomes large, we may not actually make a large number of observations.

Recall that two probability measures P and Q are said to be mutually singular (written $P \perp Q$) if there is a measurable set E such that $P(E) = 0$ and $Q(E^c) = 0$. P is absolutely continuous with respect to Q ($P \ll Q$) if $Q(E) = 0$ implies that $P(E) = 0$; and P and Q are equivalent ($P \equiv Q$) if $P \ll Q$ and $Q \ll P$.

Example: Singularity and Consistency

Let $X_1, X, \ldots$ be iid observations from the Bernoulli distribution with parameter p_1 under P, and iid Bernoulli with parameter p_2 under Q, where $p_1 \neq p_2$. By the strong law of large numbers $\bar{X}_n = (1/n)\sum_{i=1}^n X_i$ converges a.s. $[P]$ to p_1 and a.s. $[Q]$ to p_2. Let $E = \{|\lim \bar{X}_n - p_2| < \epsilon\}$, where $\epsilon < \frac{1}{2}|p_2 - p_1|$. Then $P(E) = 0$ and $Q(E^c) = 0$. In other words, we can use $\bar{X}_n$ to tell the two measures apart asymptotically. This is another way of stating that $\bar{X}$ is a consistent estimator for the success probability in a Bernoulli distribution. □

Lemma 1.3: Suppose that the offspring distributions P and Q are equivalent (i.e., $p_k > 0$ if and only if $q_k > 0$) and have $p_0 > 0$ and $q_0 > 0$. Then the restrictions to the extinction set, $\mathbf{P}|_B$ and $\mathbf{Q}|_B$, of the corresponding BGW processes are equivalent measures.

Proof: Write $B = \bigcup \{\omega\colon Z_0(\omega) = 1, \quad Z_1(\omega) = z_1, \ldots, Z_n(\omega) = z_n, Z_{n+1}(\omega) = 0\}$, where the union is over positive integers $z_1, \ldots, z_n$ and $n \geq 1$. Clearly the union is countable. Let α be a particular path in B, and assume that $\mathbf{P}(\alpha) > 0$. Write

$$\mathbf{P}(\alpha) = \prod_{j=0}^{n-1} p(z_{j+1}|z_j)\,.$$

If $p(k|l) > 0$ then there are integers $r_1, \ldots, r_l$ with $p(r_i) > 0$ and $r_1 + \cdots + r_l = k$. By assumption we also have $q(r_i) > 0$, $i = 1, \ldots, l$, so

$$q(k|l) \geq \prod_{i=1}^{l} q(r_i) > 0\,.$$

Thus

$$\mathbf{Q}(\alpha) = \prod_{j=0}^{n-1} q(z_{j+1}|z_j) > 0\,.$$ □

It will be technically convenient to introduce a slightly different description of consistency. A sequence of estimators T_n is usually called strongly consistent if $T_n \to \theta$ a.s. for all θ in the parameter space. We will regard θ as a functional on probability measures, such as the mean, and write it $\theta(P)$, for P an appropriate probability measure. Thinking momentarily of the iid case, letting P be a one-dimensional measure, a countably infinite sequence of iid random variables with marginal distribution P has a distribution which is the product measure $\mathbf{P} = \Pi_1^\infty P$. Fisher (1922b) introduced a finite sample concept of consistency that simply says that from iid observations $x_1, \ldots, x_n$ from the distribution P, the parameter $\theta(P)$ should be estimated by $\theta(P_n)$, where P_n is the empirical probability measure putting mass $1/n$ at each observation x_i. This is called Fisher consistency. In some cases this can be thought of as continuity of the functional θ in some appropriate topology, since $P_n \rightarrow P$, uniformly almost surely (so the topology in question is generated by the Kolmogorov distance). In this setting we can think of the estimator T as a random variable on the product space, rather than as a sequence of estimators. We call T *consistent for* $\theta(P)$ in a class Π of marginal measures if $\mathbf{P}\{T=\theta(P)\}=1$ for all $P \in \Pi$. This is a natural infinite-dimensional analog of the finite-dimensional concept of Fisher consistency.

The definition of consistency given above is easily adapted to the branching process case. We will need the more general concept of consistency on a subset of the parameter space. Call an estimator T (i.e., an $\mathcal{F}^Z$-measurable real valued function on Ω) of a parameter $\theta(P)$ *consistent on* $C \in \mathcal{F}^Z$ if for a class Π of branching process offspring probabilities we have

$$\mathbf{P}\{T=\theta(P), C\}=\mathbf{P}\{C\} \qquad \text{for all } P \in \Pi .$$

Notice in this definition that P is an offspring probability (for which we want to determine some property), whereas $\mathbf{P}$ is the corresponding BGW process measure. Each P generates a $\mathbf{P}$ according to the construction in section 2.

Theorem 1.2: Let $\Pi \subset \{P\colon p_0>0\}$. If T is consistent on B, then $\theta(P)=\theta(Q)$ for any P, Q in Π.

Proof: Assume that $\theta(P) \neq \theta(Q)$. Then by the consistency of T

$$\mathbf{P}\{T=\theta(Q), B\}=0 .$$

By Lemma 1.3 it follows that

$$\mathbf{Q}\{T=\theta(Q), B\}=0 .$$

But by the consistency of T we must have

$$\mathbf{Q}\{T=\theta(Q), B\}=\mathbf{Q}(B)>0 .$$

This is a contradiction. □

Example: An Estimate of the Offspring Mean

Suppose we want to estimate the offspring mean. Equation (1) suggests that an estimator at the nth generation may be

$$T_n=\begin{cases} Z_n/Z_{n-1}\,, & \text{if } Z_n>0\,,\\ 1\,, & \text{if } Z_n=0\,. \end{cases}$$

Let $T=\lim T_n$. By Proposition 1.2, $T=1$ on B, since $p_0>0$. Hence T is consistent for $\theta(P)\equiv 1$ on B. This is not a very interesting parameter, and does not help in estimating the offspring mean. □

The meaning of the theorem is that there are no interesting consistently estimable functionals on members of Π. The proof illustrates how closely connected the notions of consistency and of singularity of probability measures are. If two measures are singular, we can tell them apart using a consistent estimator, and vice versa. A fundamental difference between the branching process case and the iid model is the structure of the tail σ-field. In the iid case this is trivial (Kolomogorov's 0-1-law), but for a branching process it contains the extinction set B, which we have seen has nonzero probability whenever $p_0 > 0$. Thus tail-measurable random variables will not in general be trivial. We will see examples in this in the next chapter. Since consistency of a sequence of estimators is a tail property, it can only be achieved in models that have trivial tail σ-fields. Such models are often called *ergodic*. Inference for BGW processes is an instance of inference for nonergodic processes. The ramifications of this will become clear later.

5. CONSISTENTLY ESTIMABLE FUNCTIONALS

In this section we will discuss the results of Lockhart (1982) about what functionals of the offspring distribution can be consistently estimated on the explosion set $A = B^c$. The idea is simple: since successive generation sizes are the sums of increasing numbers of iid summands, a central limit effect ought to take over, enabling us to estimate only two moments. The technique used to prove this is a rather precise local limit theorem, coupled with an estimate of the rate of growth of the process.

Let Q be the distribution of a nonnegative integer valued random variable. The *lattice offset* r of Q is the smallest integer such that $Q(r) > 0$, and the *lattice size* (or lattice span) h of Q is the largest integer for which $U \equiv r (\text{mod } h) Q$ a.s., where U has the distribution Q. Dubuc and Seneta (1976) call such a Q *of type* (h, r). Let the distribution of a process starting with n individuals be denoted $\mathbf{P}^n$. This is consistent with standard notation, since the family trees of the initial individuals are independent, so the joint measure is a product measure. We write $\|\cdot\|$ for total variation norm, i.e.,

$$\|\mathbf{P}_1 - \mathbf{P}_2\| = \sup_F |\mathbf{P}_1(F) - \mathbf{P}_2(F)| .$$

Let f_i be the density of $\mathbf{P}_i$ with respect to some dominating measure $\mathbf{Q}$ (e.g., $\mathbf{P}_1 + \mathbf{P}_2$). Then, by letting $F = \{f_1 > f_2\}$ and noting that

$$\int_F (f_1 - f_2)\, d\mathbf{Q} = -\int_{F^c} (f_1 - f_2)\, d\mathbf{Q}$$

we see that

$$2\|\mathbf{P}_1 - \mathbf{P}_2\| = \int_\Omega |f_1 - f_2|\, d\mathbf{Q} . \tag{7}$$

Proposition 1.4: Let P and Q be two offspring distributions with the same mean $m>1$, the same variance, lattice size and offset. Assume also that for both distributions the quantity $\mathbf{E}X^3$ is finite (although it may be different for the two distributions). Then

$$\|\mathbf{P}^n - \mathbf{Q}^n\| \to 0 \text{ as } n \to \infty .$$

Proof: Let $\mathbf{P}^{n|k}$ denote the restriction of $\mathbf{P}^n$ to the history $\mathscr{F}_k = \sigma\{Z_j, 0 \le j \le k\}$. Clearly

$$\|\mathbf{P}^n - \mathbf{Q}^n\| = \lim_{k\to\infty} \|\mathbf{P}^{n|k} - \mathbf{Q}^{n|k}\| .$$

Let

$$l_{n,k} = \sum \left| \prod_{j=1}^{k} p(z_j|z_{j-1}) - \prod_{j=1}^{k} q(z_j|z_{j-1}) \right| ,$$

where $z_0 = n$ and the sum is over all sequences $(z_1, \ldots, z_k)$ of length k. By using (7) we see that

$$\|\mathbf{P}^{n|k} - \mathbf{Q}^{n|k}\| = \tfrac{1}{2} l_{n,k} .$$

Our task, therefore, will be to bound $l_{n,k}$.

Let κ be a fixed number with $1<\kappa<m$. Define

$$C_k = \{(z_1, \ldots, z_k)\colon z_j \ge \kappa z_{j-1},\ j = 1, \ldots, k\}$$

and

$$d_{n,k} = \sum_{C_k} \left| \prod_{j=1}^{k} p(z_j|z_{j-1}) - \prod_{j=1}^{k} q(z_j|z_{j-1}) \right| .$$

An application of the triangle inequality shows that

$$l_{n,k} \le d_{n,k} + \mathbf{P}^n(C_k^c) + \mathbf{Q}^n(C_k^c) . \tag{8}$$

The strategy is to show that $d_{n,k}$ is small, and that C_k, the set of paths which grow at least as fast as κ^k, is a very likely set for large n. We first compute an estimate of $d_{n,k}$. Notice that

$$\begin{aligned} d_{n,k} \le {} & \sum_{C_k} \left| \prod_{j=1}^{k-1} p(z_j|z_{j-1})[p(z_k|z_{k-1}) - q(z_k|z_{k-1})] \right| \\ & + \sum_{C_k} \left| q(z_k|z_{k-1}) \left[\prod_{j=1}^{k-1} p(z_j|z_{j-1}) - \prod_{j=1}^{k-1} q(z_j|z_{j-1}) \right] \right| . \end{aligned} \tag{9}$$

On C_k we have $z_{k-1} \ge \kappa^{k-1} n$. We now use the following local limit theorem from Petrov (1964).

Lemma 1.4: Let $\xi_1, \xi_2, \ldots$ be iid discrete random variables with lattice size h and offset r. Assume that $\mathbf{E}|\xi_1|^3 < \infty$. Let

$$f_n(N) = \mathbf{P}\left(\sum_{j=1}^{n} \xi_j = nr + Nh\right).$$

Then for any $p > 1/3$

$$\sum_{N=-\infty}^{N=\infty} \left| f_n(N) - \frac{h}{\sigma n^{1/2}} \phi\left(\frac{nr + Nh - nm}{\sigma n^{1/2}}\right)\right|^p = O(n^{-(p-1)/2}),$$

where $m = \mathbf{E}\xi_1$, $\sigma^2 = \mathbf{Var}\ \xi_1$, and ϕ is the normal density.

It follows from the lemma that

$$\begin{aligned} l_{n,1} &= \sum_{k=0}^{\infty} |p(k|n) - q(k|n)| \\ &\leq \sum_{N=0}^{\infty} \left| p(nr + Nh|n) - \frac{h}{\sigma n^{1/2}} \phi\left(\frac{nr + Nh - nm}{\sigma n^{1/2}}\right)\right| \\ &\quad + \sum_{N=0}^{\infty} \left| q(nr + Nh|n) - \frac{h}{\sigma n^{1/2}} \phi\left(\frac{nr + Nh - nm}{\sigma n^{1/2}}\right)\right| \qquad (10) \\ &\leq cn^{-1/2}. \end{aligned}$$

By summing the first term on the right-hand side of (9) over z_k we can bound it by

$$\sum_{C_k} \prod_{j=1}^{k-1} p(z_j|z_{j-1}) \frac{c}{n^{1/2}\kappa^{k/2}} \leq \frac{c}{n^{1/2}\kappa^{k/2}}.$$

Summing the second term on the right-hand side of (9) over z_k, ignoring the restriction that $z_k \geq \kappa z_{k-1}$ (which only makes the sum larger), we see that it is bounded by $d_{n,k-1}$. Thus

$$d_{n,k} \leq d_{n,k-1} + \frac{c}{n^{1/2}\kappa^{k/2}},$$

so

$$\sum_{j=2}^{k} (d_{n,j} - d_{n,j-1}) \leq \frac{c}{n^{1/2}} \frac{\kappa^{1/2}}{\kappa^{1/2} - 1}.$$

Since this sum telescopes, we get

$$d_{n,k} \leq d_{n,1} + \frac{c}{n^{1/2}(1 - \kappa^{-1/2})}.$$

Using (10) we see that $d_{n,1} \to 0$ as $n \to \infty$.

It remains to deal with the last two terms on the right-hand side of (8). First notice that

$$C_k \downarrow C = \{(z_0, z_1, \ldots): z_{j+1}/z_j \geq \kappa,\ j = 0, 1, 2, \ldots\}$$

as $k \to \infty$. Now compute

$$\mathbf{P}^n(C_k) = \mathbf{E}^n(C_{k-1}\mathbf{P}^{Z_{k-1}}(C_1)),$$

where $\mathbf{E}^n$ is expectation will respect to $\mathbf{P}^n$, and we used the Markov property. Using Chebyshev's inequality we see that

$$\mathbf{P}^n(C_1^c) = \mathbf{P}^n(Z_1 \leq \kappa n) \leq \mathbf{P}^n(|Z_1 - n\mu| \geq n(\mu - \kappa)) \leq \left(\frac{\sigma}{m - \kappa}\right)^2 \frac{1}{n},$$

so

$$\mathbf{P}^n(C_k) \geq \mathbf{E}^n\left(C_{k-1}\left(1 - \left(\frac{\sigma}{m - \kappa}\right)^2 \frac{1}{Z_{k-1}}\right)\right).$$

On C_{k-1}, $Z_{k-1} > n\kappa^{k-1}$, so

$$\mathbf{P}^n(C_k) \geq \mathbf{P}^n(C_{k-1})\left(1 - \left(\frac{\sigma}{m - \kappa}\right)^2 \frac{1}{n\kappa^{k-1}}\right).$$

Since $\mathbf{P}^n(C_1) > 0$ we conclude that

$$\mathbf{P}^n(C_k) \geq \mathbf{P}^n(C_1) \prod_{j=1}^{k-1} \left(1 - \left(\frac{\sigma}{m - \kappa}\right)^2 \frac{1}{n\kappa^j}\right).$$

Since for $0 < x < 1$, $\log(1 - x) > -x/(1 - x)$, we can bound the product below for $n > \sigma^2/(m - \kappa)^2$ by

$$k_n = \exp\left(-\frac{c}{n - c}\,\frac{1}{\kappa - 1}\right),$$

where $c = \sigma^2/(m - \kappa)^2$. Since k_n does not depend on k, we let $k \to \infty$ and obtain the bound

$$\mathbf{P}^n(C) \geq \mathbf{P}^n(C_1)k_n,$$

and since $k_n \to 1$ and $\mathbf{P}^n(C_1) \to 1$ we see that

$$\lim_{n \to \infty} \mathbf{P}^n(C) = 1.$$

The same computation for $\mathbf{Q}^n$ verifies that the right-hand side of (8) converges to zero as $n \to \infty$, so that $\|\mathbf{P}^n - \mathbf{Q}^n\| \to 0$. □

We are now able to decide what functionals are consistently estimable on the explosion set A. We say that two offspring distribution P and Q are *not singular on* $\{n \geq 2\}$ if there is an $m \geq 2$ such that $P(\{m\}) > 0$ and $Q(\{m\}) > 0$.

Theorem 1.3: Suppose T is a consistent estimator on A of $\theta(P)$ for $P \in \Pi$, a class of branching process offspring probabilities. Then, whenever P and Q in Π have the same mean, variance, lattice size and offset, are not singular on $\{n \geq 2\}$ and have $\mathbf{E}X^3$ finite, we have $\theta(P) = \theta(Q)$.

Proof: Suppose the conclusion does not hold, and fix P and Q in Π such that $\theta(P) \neq \theta(Q)$, P and Q satisfy the hypotheses of Proposition 4 and are not singular on $\{n \geq 2\}$. Let $m \geq 2$ be such that $\min(P(\{m\}), Q(\{m\})) > 0$, and choose N so that for $n \geq N$ we have $\|\mathbf{P}^n - \mathbf{Q}^n\| < 1/4$ and $\min(\mathbf{P}^n(A), \mathbf{Q}^n(A)) > 3/4$. Choose k such that $n_0 m^k \geq N$. Notice that

$$\mathbf{P}^{n_0}(T = \theta(Q), A) = 0$$

and

$$\mathbf{Q}^{n_0}(T = \theta(Q), A) = \mathbf{Q}^{n_0}(A) > 0$$

by the consistency assumption. Hence $\mathbf{P}^{n_0}$ and $\mathbf{Q}^{n_0}$ are singular on A, consequently on $A \cap D$, where $D = \{Z_j = n_0 m^j,\ j = 1, \ldots, k\}$. But $A \cap D$ has positive probability for both $\mathbf{P}^{n_0}$ and Q^{n_0}, so $\mathbf{P}^{n_0 m^k}$ and $\mathbf{Q}^{n_0 m^k}$ are singular on A, which contradicts the choice of N above. □

Theorem 1.3 tells us what parameters we *may* be able to estimate: offspring mean, variance, lattice size, and lattice offset. It does not say that these parameters in fact are estimable. This question will be pursued in the next two chapters.

Remarks:

(i) If we are interested in weak consistency, Theorem 1.3 may be applied to a strongly convergent subsequence.

(ii) Exercise 11 shows that the lattice offset is not in general estimable on the explosion set. On the other hand, if we can estimate the lattice size consistently when starting from one ancestor, we can read off the lattice offset from Z_1 when $Z_0 = 1$. We will return to this question in chapter 3.

(iii) The condition $\mathbf{E}X^3 < \infty$ is not necessary to get a rate of convergence in the local central limit theorem. Using results of Hall (1982, Theorem 5.4) one can improve the condition in Proposition 1.4 to $\mathbf{E}X^2 \log^{\alpha}(X+3) < \infty$ for some $\alpha > \frac{1}{2}$. However, the local limit theorem does not yield a rate of convergence if we only assume a finite variance. Such a rate is needed for the technique of proof used here. □

6. OTHER SAMPLING SCHEMES

It should be fairly clear (you may want to prove it) that if we observe the entire family tree, we can estimate the offspring the distribution consistently, provided that $p_0 = 0$ so that extinction is excluded. That leads to the question whether there may be intermediate cases, such that more parameters than in the generation size case can be estimated conditionally consistently, and yet one does not observe the entire family tree. One such intermediate scheme is to randomly sample individuals in the process, and determine each sampled individuals family size. Obviously, this scheme yields no information about the probability of zero offspring.

Example: Using Birth Order to Estimate Family Sizes

Burks (1933) discusses the demographic problem of obtaining data on the distribution of family sizes. Taking a random sample of individuals and ascertaining their family sizes yields a sample which is biased towards large family sizes. To deal with this problem, she suggests taking a random sample of individuals from the population, and ascertaining their birth order rather than their family size. One way this used to be achieved was by sampling birth certificates, which in the past (in the United States) usually required that mothers report whether a particular birth was their first, second, third, and so on. The number of individuals in the sample with birth order k should be approximately $n \sum_{j \geq k} p_j$, where n is the sample size, since given that the eventual family size is j, the probability is $1/j$ for each particular birth order entering the sample, and a family of size j has probability jp_j to enter the sample. This assumes that the poplation behaves in a stationary fashion. We will return to this example in the next chapter. □

Waugh (1981) and Joffe and Waugh (1982) have developed some theory for the incomplete observational scheme where individuals are sampled from one or several consecutive generations, and their respective familys sizes are recorded. Maki and McDunnough (1989a) have studied the statistical problems involved. Their main structural result is that family sizes are asymptotically iid, with asymptotic distribution (jp_j/m). In this section we will sketch their proof, and illustrate the use of the theory to questions of what can be estimated under various circumstances.

Much of the difficulty in looking at samples from a generation arises from the fact that two or more individuals may belong to the same family. Maki and McDunnough (1989a) show that this event has a probability that goes to zero exponentially fast, provided that $r_n^2/m^n \to 0$, where r_n is the number of individuals sampled in the nth generation. The argument for the case where the offspring distribution is of bounded support is outlined in Exercise 12. This enables approximating the joint distribution of sampled family sizes by the distribution of family sizes restricted to come from different families. Denote the r_n sampled family sizes from the nth generation by $\mathbf{Y}_n$, and the set {all selected individuals from different families} by D_n. We have $\mathbf{P}(D_n) \to 1$. For simplicity we will assume that $p_0 = 0$ and $p_1 < 1$, so that the process is explosive. This is a natural assumption when we sample from a generation, and cannot observe zero offspring. We will assume that $r_n \le Z_n$; this is certain for $r_n \le n$, and highly likely if r_n is much smaller than m^n and n is large. By first conditioning on the unsampled family sizes it is easy to verify (Exercise 13) that

$$\mathbf{P}\{\mathbf{Y}_n = \mathbf{i}_n, D_n\} = \left(\prod_{j=1}^{r_n} i_{jn} p_{i_{jn}}\right) \mathbf{E}\left[\binom{Z_{n-1}}{r_n}\Big/\binom{\mathbf{i}_n^T \mathbf{1} + X_{r_n+1,n} + \cdots + X_{Z_{n-1},n}}{r_n}\right] \tag{11}$$

where $\mathbf{i}_n^T = (i_{1n}, \ldots, i_{r_n n})$ and $\mathbf{1}$ is a vector of r_n ones. Let $\tilde{c}_n(\mathbf{t}_n) = \mathbf{E}(1_{D_n} \exp[i\mathbf{t}_n^T \mathbf{Y}_n])$ where $\mathbf{t}_n$ is a vector of length r_n. Using (11), we see that

$$\tilde{c}_n(\mathbf{t}_n) = \mathbf{E}\left[X_{1,n} \ldots X_{r_n,n} \binom{Z_{n-1}}{r_n}\binom{Z_n}{r_n}^{-1} \exp(i\mathbf{t}_n^T \mathbf{X}_n)\right],$$

where $\mathbf{X}_n = (X_{1,n}, \ldots, X_{r_n,n})^T$. We are really interested in $c_n(\mathbf{t}_n) = \mathbf{E}(\exp[i\mathbf{t}_n^T \mathbf{Y}_n])$, i.e., the joint characteristic function of the sampled variates. Let $c(\mathbf{t}_n)$ be the joint characteristic function of r_n iid random variables from the distribution (jp_j/m).

Proposition 1.5: Suppose that $\sigma^2 < \infty$ and that $r_n^2/m^n \to 0$. Then

$$\lim_{n\to\infty} |c_n(\mathbf{t}_n) - c(\mathbf{t}_n)| = 0 .$$

Proof: First note that $c(\mathbf{t}_n) = \mathbf{E}[m^{-r_n} X_{1,n} \ldots X_{r_n,n} \exp(i\mathbf{t}_n^T \mathbf{X}_n)]$. Estimate

$$|c_n(\mathbf{t}_n) - c(\mathbf{t}_n)| \le |c_n(\mathbf{t}_n) - \tilde{c}_n(\mathbf{t}_n)| + |\tilde{c}_n(\mathbf{t}_n) - c(\mathbf{t}_n)| .$$

The first quantity on the right-hand side is bounded by $1 - \mathbf{P}(D_n)$ which goes to zero. The second term is bounded by

$$\mathbf{E}\left[X_{1,n} \ldots X_{r_n,n} \left|\binom{Z_{n-1}}{r_n}\binom{Z_n}{r_n}^{-1} - m^{-r_n}\right|\right].$$

To handle this quantity, we first note that the product $Z_{1,n} \ldots X_{r_n,n}$ can be replaced by any other product of r_n distinct $X_{j,n}$. Hence the expectation has the same value if we replace this product by the average over all such products. Denoting these products Π_j, and setting $C_n = \binom{Z_n}{r_n}$, we need to study the convergence of

$$\mathbf{E}\left[m^{-r_n} C_n^{-1} \sum_1^{C_n} \Pi_j \left| m^{r_n} \binom{Z_{n-1}}{r_n}\binom{Z_n}{r_n}^{-1} - 1\right|\right].$$

The first factor in this expression converges in probability to 1, by Chebyshev's inequality. We shall argue that the second term goes in probability to zero. Since both terms have bounded expectation, the required L_1-convergence will then follow. Write

$$\left| m^{r_n} \binom{Z_{n-1}}{r_n} \binom{Z_n}{r_n}^{-1} - 1 \right| = \left| \left(\frac{mZ_{n-1}}{Z_n} \right)^{r_n} \prod_{j=1}^{r_n - 1} \frac{1 - j/Z_{n-1}}{1 - j/Z_n} - 1 \right|$$

We bound the product above by $(1 - r_{n-1}/Z_n)^{r_n}$, and below by $(1 - r_{n-1}/Z_{n-1})^{r_n}$. Using Theorem 1.1(ii) we see that $Z_{n-1} = m^{n-1}W + O_{\mathbf{P}}(m^{n-1})$ whence the product converges in probability to 1. It remains to show that $(mZ_{n-1}/Z_n)^{r_n}$ converges in probability to 1. Let $\bar{m}_n = Z_n/Z_{n-1}$. Then

$$\left(\frac{mZ_{n-1}}{Z_n} \right)^{r_n} = \exp\left(r_n \log \frac{m}{\bar{m}_n} \right) = \exp\left[-r_n \frac{\bar{m}_n - m}{m} + O(r_n(\bar{m}_n - m)^2) \right].$$

In the next chapter we will demonstrate (Theorem 2.4) that $\bar{m}_n - m = O_{\mathbf{P}}(Z_{n-1}^{1/2})$. Using Theorem 1.1(ii) again see that

$$\left(\frac{mZ_{n-1}}{Z_n} \right)^{r_n} = \exp(O_{\mathbf{P}}(r_n m^{n/2})) \rightarrow 1 \text{ in probability.} \qquad \square$$

The result has a straightforward extension to sampling several generations. Since the asymptotic distribution is that of iid random variables, we can estimate the distribution $(q_j) \equiv (jp_j/m)$ consistently from $\mathbf{Y}_n$. Since $p_j = (q_j/j)/(\Sigma\, q_k/k)$, we can reconstruct the offspring distribution from the sampled family sizes in a consistent fashion (remember that we are assuming that $p_0 = 0$).

Suppose that instead of observing the individual family sizes in the sample, we only get to see their sum, i.e., $\mathbf{Y}_n^T\mathbf{1}$. Using the theory we just sketched, the asymptotic distribution of the sum is that of a sum of iid random variables, in other words a sampled random walk. Inference from such observations was studied by Guttorp and Siegel (1985), who showed that the kth moment of the underlying distribution can be estimated if the r_n grow slowly, more precisely if $\Sigma\, r_n^{-(k-2)}$ is infinite. For $k = 3$ this was shown to be both necessary and sufficient for consistent estimation of the kth moment. A more general result towards a necessary condition was developed by Maki and McDunnough (1989b), who showed, using methods very similar to those in the previous section, that when $\Sigma\, r_n^{-(k-2)/2} < \infty$, moments of order greater than k cannot be consistently estimated. Unfortunately, this condition is not necessary even at $k = 3$. Guttorp and Lockhart (1989) subsequently showed that the Guttorp and Siegel condition is necessary for all k.

7. SOURCES

Section 1 is largely based on Albertsen (1976), with additional material from Heyde and Seneta (1977, section 5.9), Jagers (1975), and Kendall (1966), Keiding (personal communication) provided information about the Krarup family tree.

Proposition 1.1 is a special case of a general construction (Theorem VI.5.1) from Harris (1963). The construction is originally due to Otter (1949). Lemma 1.1 is Jager's (1975) Lemma 2.1.1.

The proof of Proposition 1.2 is largely that given by Christensen (see Albertsen and Kristensen, 1976). Proposition 1.3 is taken from Jagers (1975, Theorem 2.3.2). The Application following Proposition 1.3 is based on some computations in Kendall (1975) and Pollard (1973), modified by the revised fit. The results in Theorem 1.1 date at least back to Yaglom (1947), who proved convergence in distribution. Hawkins and Ulam obtained the same reslt in an unpublished technical report from Los Alamos in 1944. Harris (1948) proved L_2-convergence, and Doob suggested the martingale approach. The functional equation in Theorem 1.1(v) was analyzed by Poincaré (1890).

The proofs of Lemma 1.3, Proposition 1.4, Theorems 1.2 and 1.3 are all variants on the work of Lockhart (1982). Similar results have been developed in slightly different contexts by Maki and McDunnough (1989b) and Winnicki (1988).

Proposition 1.5 is Maki and McDunnough (1989a), Theorem 2.

EXERCISES

1. Prove that $P_k(s) = P(P_{k-1}(s))$ provided that $z_0 = 1$.

2. (a) Compute $\mathbf{E}Z_n^3$ and $\mathbf{E}Z_n^4$, assuming that P has four finite moments.
   (b) Let $Y_n = \Sigma_{i=0}^n Z_i$. Show that

$$\mathbf{E}Y_n = \frac{m^{n+1} - 1}{m - 1}$$

and that

$$\mathbf{Var}\ Y_n = \frac{\sigma^2}{(m-1)^2}\left(\frac{m^{2n+1} - 1}{m - 1} - (2n+1)m^n\right).$$

Hint: Let $\alpha_n = \mathbf{E}Y_n(Y_n - 1)$, and show that α_n satisfies the difference equation

$$\alpha_{n+1} = 2m\mathbf{E}Y_n + (\sigma^2 + m(m-1))\mathbf{E}^2Y_n + m\alpha_n .$$

3. Let $p_k = bc^{k-1}$, $k \geq 1$ and $p_0 = 1 - \Sigma_{k\geq 1} p_k$.

(a) Write $P(s)$ in the form

$$P(s) = \frac{\alpha + \beta s}{\phi + \delta s}, \qquad \alpha\delta - \beta\gamma \neq 0$$

This is called a *fractional linear* generating function.

(b) Show that $P_2(s)$ is a fractional linear generating function.

(c) Let s_0 be the nonnegative root to $P(s) = s$ different from 1. Compute

$$P_n(s) = 1 - m^n\left(\frac{1-s_0}{m^n - s_0}\right) + m^n \frac{\left(\dfrac{1-s_0}{m^n - s_0}\right)^2 s}{1 - \left(\dfrac{m^n - 1}{m^n \quad s_0}\right) s}$$

where $m = p'(1) \neq 1$.

(d) Let T be the time to extinction (i.e., the first n with $Z_n = 0$). Find $\mathbf{P}\{T = n\}$.
Hint: $\{T = n\} = \{Z_n = 0, Z_{n-1} > 0\}$.

4. We say that the family $I(\omega)$ contains an *infinite line of descent* if there are positive integers $i_1, i_2, \ldots$ such that

$$(i_1 i_2 \ldots i_k) \in I_k(\omega),\ k = 1, 2, \ldots.$$

Prove that $I(\omega)$ contains an infinite line of descent if and only if all generations $I_1(\omega), I_2(\omega), \ldots$ are nonempty.

5. Define $\omega_i = (\nu_i, \nu_{i1}, \nu_{i2}, \nu_{i11}, \ldots)$, the family history of (i). Show that $Z_k(\omega) = \Sigma_{i=1}^{\nu_0(\omega)} Z_{k-1}(\omega_i)$, provided $\nu_0(\omega) > 0$.

6. Let $(S_n, n \geq 0)$ be a random walk. Define stopping times $T_{-1} = 0$, and

$$T_n = \sum_{k=1}^{n} Z_k,\ n = 0, 1, 2, \ldots,$$

where

$$Z_{n+1} = S_{T_n} - S_{T_{n-1}}.$$

Show that $(Z_n, n \geq 0)$ is a BGW process.

7. Demonstrate the following properties of $P_k(s)$:

(a) $P_k(s) \uparrow q$ for $0 \leq s < q$ as $k \to \infty$.

(b) $P_k(s) \downarrow q$ for $q < s < 1$ as $k \to \infty$.

(c) If $s = q$ or 1 then $P_k(s) = s$ for all k.
(d) The convergence in (a) is uniform.

8. (a) Assume that P is a Poisson distribution with mean $m > 1$. Verify that if m is near 1, $q \approx (m - 1/m)^2 + m^{-2}$.
(b) By a Taylor expansion of $P(s)$ around 1, show that for any supercritical offspring distribution with mean near 1, $1 - q \approx 2(m-1)/[\sigma^2 + m(m-1)]$.
(c) Show that for q near 1, $q \approx \exp(-2(m-1)/\sigma^2)$.

9. Prove that the solution to the equation $\phi(ms) = P(\phi(s))$ of Theorem 1.1(v), subject to $\phi'(0) = 1$, is unique.
Hint: Let $\psi(s) = \phi(-is)$. If ψ_1 and ψ_2 both satisfy the equation, write $\psi_1(t) - \psi_2(t) = t\gamma(t)$, where γ is continuous with $\gamma(0) = 0$. Show that $|\gamma(t)| \geq |\gamma(mt)|$.

10. Prove that if P and Q satisfy the hypotheses of Proposition 1.4 and $P \ll Q$, then for any n_0, $\mathbf{P}^{n_0} \ll \mathbf{Q}^{n_0}$.

11. Let P be supported on $\{1, 3\}$ with probabilities $\frac{1}{2}$ each, and Q on $\{0, 2, 4\}$ with probabilities 1/8, 6/8 and 1/8.
(a) Show that $\mathbf{P}^2 \ll \mathbf{Q}^2$.
(b) Deduce that the lattice offset is not estimable in general.

12. Suppose that the support of the offspring distribution is $\{l, \ldots, k\}$. Let

$$D_n = \{\text{all } r_n \text{ selected individuals belong to the same family}\} .$$

Show that

$$k^{r_n} \binom{Z_{n-1}}{r_n} \binom{kZ_{n-1}}{r_n}^{-1} \leq \mathbf{P}(D_n | Z_{n-1}) \leq l^{r_n} \binom{Z_{n-1}}{r_n} \binom{lZ_{n-1}}{r_n}^{-1} \leq 1$$

and deduce that if $r_n^2/m^n \to 0$ we have that $\mathbf{P}(D_n) \to 0$.

13. Verify formula (10).
Hint: Condition on the unsampled family sizes first.

14. (a) Show that a BGW process (Z_n), conditional upon nonextinction, is a Markov chain with n-step transition probabilities

$$q_{ij}^{(n)} = \frac{1 - q^j}{1 - q^i} \, p_{ij}^{(n)}$$

where $p_{ij}^{(n)}$ are the n-step transition probabilities for (Z_n).

(b) Let $T_k^{(i)} = \inf\{n: Z_n > k \mid Z_0 = i\}$ be the first passage time of k, starting from i ancestors. Show that

$$\mathbf{P}(T_k^{(i)} \leq N) = 1 - q_i(k, N)$$

where $q_i(k, N)$ is the ith element of $\mathbf{Q}_k^N \mathbf{1}$, where $\mathbf{Q}_k = (q_{ij}; 1 \leq i, j \leq k)$.

(c) Find the mean time to exceed 6 offspring (conditional upon nonextinction) for the US 1920 white male population (Application in section 1.2).

CHAPTER 2

Estimating the Offspring Mean

The problem of estimating the offspring mean of a BGW process has been quite intensively studied. Because of the simplicity of the model, many interesting features that make inference for stochastic processes different from classical statistical inference can be highlighted. We will present some different estimators, and compute some of their properties, both for finite samples and asymptotically. The classical theory for exponential families of distributions can be adapted to nonergodic models as well, and concepts such as efficiency can be studied. In order to develop a theoretical foundation for conditional inference in stochastic process problems, we discuss ancillarity and curvature. We also look at asymptotic testing, where again the theory is quite different from the iid case.

1. AN EXAMPLE

The problem of determining the fraction of a community that must be vaccinated in order to prevent major epidemics of a communicable disease is a crucial public health problem. In order to describe an epidemic, the population is divided into three possible health states. An individual can be susceptible to infection by a given disease agent, he may have been infected by the agent and be infectious (possibly after an incubation period), or he is removed from the epidemic by death, by isolation, or by immunity or other natural loss of infectiousness. Initially all members of the population are susceptible to infection. The epidemic starts when one or many infectious individuals enter the population and come into contact with its members. A susceptible person is infected if he has adequate contact with an infectious individual. General theory of epidemic models can be found in Bailey (1975), and their statistical inference is discussed, e.g., in Becker (1976) and Rida (1991). Following Bartoszyński (1967) we will see how a branching process can be used to model an epidemic process.

Assume that every individual who becomes infected first passes through an incubation period of length X, followed by a period of infectiousness of length Y. Let $F(s, t)$ be the joint pgf for (X, Y). At each day during the period of incubation an individual has a probability $1-\alpha$ of being discovered and isolated. During the period of infectiousness this probability is $1-\beta$. An infectious individual who has not previously been discovered contacts a certain number of noninfected members of the population each day. These numbers are iid according to a distribution with pgf $R(t)=\Sigma\, r_j t^j$, having a finite mean. Each contact with an infectious individual yields an infection with probability γ, independently of the results of other contacts. All individuals act independently, and independent of the history of the process.

Consider a single individual who became infected at a certain moment, and let w_n be the probability that he will be effectively infectious (infectious and not isolated) during exactly n days. He must remain undiscovered during his incubation period, and during the first n days of his infectiousness period. If $X=m$ and $Y=n$ this probability is $\alpha^m\beta^n$. If instead $Y=k>n$, the probability is $\alpha^m\beta^n(1-\beta)$. Let $Q(s, t)=\Sigma\, s^m t^n \mathbf{P}(X=m,\ Y>n)$. Then

$$w_n=\beta^n F(\alpha, 1)=(1-\beta)Q(\alpha, 1)\,.$$

For $n=0$ we have that $w_0=1-F(\alpha, 1)$. Now let ν_k be the probability of exactly k contacts leading to infection during one day of infectiousness. Then

$$\nu_k=\sum_{j\geq k}\binom{j}{k}\gamma^k(1-\gamma)^{j-k}r_j$$

so that the generating function for (ν_k) is

$$V(t)=\sum \nu_k t^k=R(\gamma t+1-\gamma)\,.$$

Thus the generating function of the joint number of infected during n days of infectiousness is $R(\gamma t+1-\gamma)^n$. The generating function for the number of direct descendants of an infected individual is

$$\begin{aligned}G(t)&=1-F(\alpha, 1)+\sum_{n\geq 1} w_n R(\gamma t+1-\gamma)^n\\&=1-F(\alpha, 1)+F(\alpha, \beta R(\gamma t+1-\gamma))\\&\quad+(1-\beta)Q(\alpha, \beta R(\gamma t+1-\gamma))\,.\end{aligned}$$

Noting that

$$Q(s, t)=\frac{F(s, 1)-F(s, t)}{1-t}\,,$$

we can write

$$G(t) = 1 - F(\alpha, 1) + \frac{(1-\beta)F(\alpha, 1) + \beta(1 - R(\gamma t + 1 - \gamma))F(\alpha, \beta R(\gamma t + 1 - \gamma))}{1 - \beta R(\gamma t + 1 - \gamma)} .$$

The offspring mean consequently is (for $\beta < 1$)

$$G'(1) = \frac{\beta}{1-\beta}\gamma R'(1)(F(\alpha, 1) - F(\alpha, \beta)) .$$

We have that $G'(1) > 1$ if and only if

$$F(\alpha, \beta) < F(\alpha, 1) - \frac{1-\beta}{\gamma r \beta}$$

where $r = R'(1)$ is the expected number of contacts per day. By considering the left- and right-hand sides of the inequality as functions of β we see that both have positive derivatives, whereas their second derivatives are of opposite sign. Let $D(\alpha) = \partial F(s, t)/\partial t|_{s=\alpha, t=1}$ be the average length of the period of infectiousness of individuals who remained undiscovered during their incubation period. Then $G'(1) > 1$ if and only if the expected number of infections $D(\alpha)\gamma r > 1$, and $\beta > \beta_0$ where β_0 is the intersection of the two functions of β, i.e., the smallest root of the equation

$$\gamma r x F(\alpha, x) - \gamma r x F(\alpha, 1) + 1 - x = 0 .$$

For the special case where the incubation period is of constant length M we get $D(\alpha) = D\alpha^M$, where D is the average length of the period of infectiousness. Then we are in the subcritical domain provided that either $\gamma r D \alpha^M \leq 1$ or $\beta \leq (1 + \gamma r \alpha^M)^{-1}$. Preventive measures can decrease α, β, or γ. Changes in social behavior would affect r.

The main problem with the Bartoszyński model is that it assumes an infinite pool of susceptibles. A BGW process can therefore only be used to approximate the infectious population during the early stages of an epidemic (Becker, 1977). Since the number of susceptible individuals decreases as the epidemic progresses, it may be unreasonable to assume that the offspring distribution is the same from generation to generation. However, for early stages of epidemics in large populations the assumptions underlying the BGW model are not bad.

In order for an epidemic to become major, a large buildup of cases is needed in the early stage. Consequently we can call an epidemic major if the offspring mean is >1, and minor if it is ≤1 (so that the extinction probability is 1). In order to prevent major epidemics it is necessary to ensure adequate vaccination in the community so as to make the offspring mean less than one. Suppose that we select a proportion θ of the population at random for vaccination. If the vaccination is effective, the offspring distribution changes to $P^*(s) = \theta + (1-\theta)P(s)$ with mean $m^* = (1-\theta)m$, so that $m^* < 1$ only if $\theta > 1 - 1/m$. For a numerical illustration, assume that the offspring distribution is Poisson. Then the probability $1-q$ of a major epidemic is a function of m:

m	1	1.05	1.1	1.4	1.8
$1-q$	0	0.09	0.18	0.51	0.73
$1-1/m$	0	0.05	0.09	0.29	0.44

The third line of the table contains the proportion of vaccination needed to bring the new mean below 1. If one accepts this model it becomes of considerable importance to be able to estimate m as accurately as possible.

2. METHOD OF MOMENT ESTIMATORS

There are a variety of possible estimators of the offspring mean of a BGW process. We will first consider some nonparametric estimators, and in a later section use parametric models. A simple method of moments estimator obtains from equating the current generation size to its expected value,

$$Z_n = \mathbf{E}Z_n = m^n ,$$

so that

$$m_n^* = Z_n^{1/n} .$$

m^* was suggested by Heyde (1975). It is immediate from Jensen's inequality that m^* will underestimate m on average. One consequence of Theorem 1.1 is that $Z_n = Wm^n + o_{\mathbf{P}}(m^n)$. Thus, if $W > 0$, we can plot $\log Z_n$ against n and should, after some initial wiggling, get essentially a straight line, provided that the path does not become extinct. Different realizations of the same process will yield lines with different intercepts but the same slope $\log m$. This is illustrated in Figure 2.1. For large n, $\log Z_n/n$ can be used to estimate this slope. Exponentiating, we get m^*.

Example: The Geometric Case

In the geometric offspring case, we use Exercise 1.3 to see that Z_n has a modified geometric distribution with fractional linear generating function

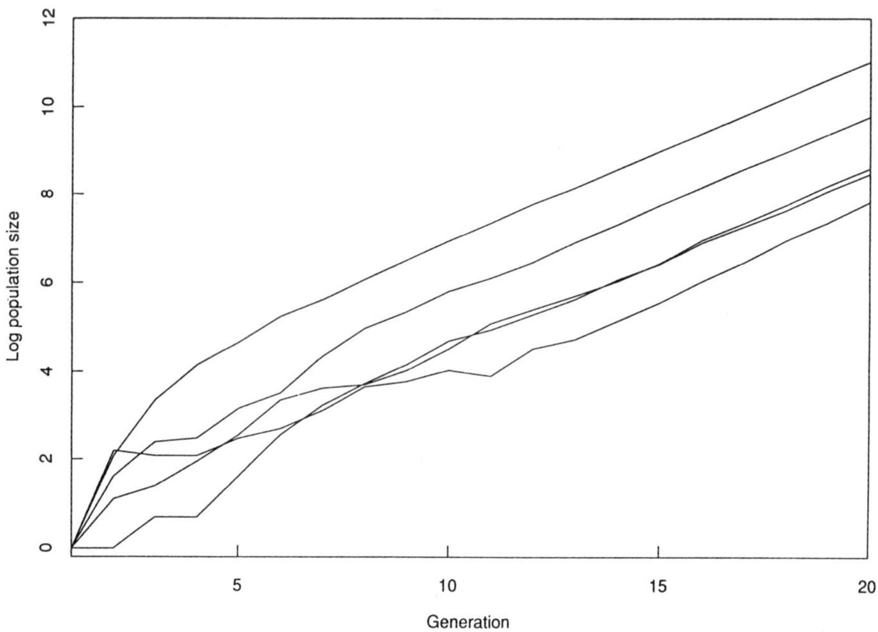

FIGURE 2.1. log Z_n plotted against n for some nonextinct paths. The offspring distribution is Poisson with mean 1.6.

$$P_n(s) = 1 - m^n\left(\frac{1-1/m}{m^n - 1/m}\right) + \frac{m^n\left(\dfrac{1-1/m}{m^n-1/m}\right)^2 s}{1 - \dfrac{m^n - 1}{m^n - 1/m}s}$$

$$\equiv \alpha_n + \frac{\beta_n s}{1-\gamma_n s}\,.$$

By writing $P_n(s) = \Sigma_0^\infty P_k^{(n)} s^k$ and identifying coefficients of s, we see that

$$p_0^{(n)} = \alpha_n\,, \qquad p_k^{(n)} = \beta_n \gamma_n^{k-1}\,.$$

Therefore $\mathbf{E}Z^{1/n} = \beta_n \Sigma_1^\infty k^{1/n}\gamma_n^{k-1}$. We approximate the sum by an integral to get

$$\mathbf{E}Z^{1/n} \approx \beta_n \gamma_n^{-1}\left(\log\frac{1}{\gamma_n}\right)^{-(1/n)-1}\Gamma\left(\frac{1}{n}+1\right).$$

For large n and $m > 1$ we approximate $\gamma_n \approx 1 - (1-1/m)m^{-n}$, and $\beta_n \approx m^{-n}(1-1/m)^2$. Joining together the pieces, using that $\Gamma(1+1/n) \approx 1$, we see that $\mathbf{E}m_n^* \approx m - 1$, with a bias of -1. For example, if $p = 0.45$ (so $m = 1.222$) and $n = 20$, the exact mean is 0.235. Conditional on $Z_n > 0$, on the other hand, all probabilities are divided by $1 - \alpha_n \approx 1 - 1/m$, whence the conditional mean is

$$\mathbf{E}(Z_n^{1/n} | Z_n > 0) \approx m . \qquad \square$$

We may also apply the method of moments to the conditional distribution of Z_n, given the previous generation sizes. Since (using the Markov property)

$$\mathbf{E}(Z_n | \mathcal{F}_{n-1}) = \mathbf{E}(Z_n | Z_{n-1}) = Z_{n-1} m ,$$

we get

$$\bar{m}_n = \begin{cases} Z_n / Z_{n-1} , & \text{if } Z_{n-1} > 0 , \\ 1 , & \text{if } Z_{n-1} = 0 , \end{cases}$$

an estimator first proposed by Lotka (1939). Some finite sample properties of $\bar{m}_n$ are easily established. We first compute the conditional moments, given nonextinction.

Proposition 2.1

$$\mathbf{E}(\bar{m}_n | Z_{n-1} > 0) = m ,$$

$$\mathbf{Var}(\bar{m}_n | Z_{n-1} > 0) = \sigma^2 \mathbf{E}(Z_{n-1}^{-1} | Z_{n-1} > 0) .$$

Proof: For any $z > 0$ we have that $\mathbf{E}(\bar{m}_n | Z_{n-1} = z) = m$ since Z_n given $Z_{n-1} = z$ is the sum of z iid random variables. Likewise we find that $\mathbf{Var}(\bar{m}_n | Z_{n-1} = z) = \sigma^2 / z$. Averaging over all such z the result follows. $\square$

In order to compute unconditional moments of $\bar{m}_n$ we need the following easy result, the proof of which is left to the reader.

Lemma 2.1: Let F be any measurable set, and X a random variable with finite variance. Then

$$\begin{aligned} \mathbf{Var}(X) = &\mathbf{Var}(X | F)\mathbf{P}(F) + \mathbf{Var}(X | F^c)\mathbf{P}(F^c) \\ &+ (\mathbf{E}(X | F) - \mathbf{E}(X | F^c))^2 \mathbf{P}(F)\mathbf{P}(F^c) . \end{aligned}$$

Proposition 2.2

$$\mathbf{E}\bar{m}_n = m\mathbf{P}(Z_{n-1} > 0) + \mathbf{P}(Z_{n-1} = 0) ,$$

$$\begin{aligned} \mathbf{Var}\, \bar{m}_n = &\sigma^2 \mathbf{E}(Z_{n-1}^{-1} | Z_{n-1} > 0)\mathbf{P}(Z_{n-1} > 0) \\ &+ (m-1)^2 \mathbf{P}(Z_{n-1} > 0)\mathbf{P}(Z_{n-1} = 0) . \end{aligned}$$

Proof: Immediate from Proposition 2.1 and Lemma 2.1. $\square$

Example: The Geometric Case

In the geometric case for large n we know that $\mathbf{P}(Z_{n-1}=0)\approx q=1/m$, so $\mathbf{E}\bar{m}_n\approx m-1+1/m$ for $m>1$. The bias is smaller than that of m_n^*. For $p=0.45$ and $n=20$ the exact mean is 1.041, whereas the approximation yields 1.040. □

Returning to Figure 2.1, we want to estimate the slope of the path on a logarithmic scale. A reasonable estimate of the slope is $\log Z_n-\log Z_{n-1}$, which when exponentiated yields $\bar{m}$. We would expect $\bar{m}$ to improve on m^*, since the variability of the path around a straight line tends to decrease at later generations.

The two estimators above do not take much account of the previous history of the process. A natural approach would be to try a weighted average of the estimators $\bar{m}_k$ or m_k^*. We will concentrate on $\bar{m}$ here, leaving m^* to the exercises. Since $\bar{m}_n$ was obtained by a conditional method of moments argument, it seems reasonable to find the best linear combination of the $\bar{m}_k$, with weights proportional to the conditional variances given Z_{k-1}. This means giving more weight to larger generations. The estimator becomes

$$\hat{m}_n=\sum_{i=1}^{n}\frac{Z_{i-1}}{Y_{n-1}}\,\bar{m}_i=\frac{Y_n-1}{Y_{n-1}}\,,$$

where $Y_k=\Sigma_{i=0}^{k}Z_i$ is the total population size up to and including the kth generation. The definition of $\hat{m}_n$ makes sense also when the population becomes extinct: on $\{Z_{n-1}=0\}$ we have that $Y_{n-1}=Y_\infty=\lim_{k\to\infty}Y_k$, so that on this set $\hat{m}_n\to 1-1/Y_\infty$. This estimator will claim a subcritical population whenever the population becomes extinct before the end of the observation period.

Proposition 2.3

$$\mathbf{E}(\hat{m}_n|Z_{n-1}>0)=1+m\mathbf{E}\left(\frac{Z_{n-1}}{Y_{n-1}}\middle|Z_{n-1}>0\right)-\mathbf{E}\left(\frac{1}{Y_{n-1}}\middle|Z_{n-1}>0\right),$$

$$\mathbf{Var}(\hat{m}_n|Z_{n-1}>0)=\sigma^2\mathbf{E}\left(\frac{Z_{n-1}}{Y_{n-1}^2}\middle|Z_{n-1}>0\right)+\mathbf{Var}\left(\frac{mZ_{n-1}-1}{Y_{n-1}}\middle|Z_{n-1}>0\right).$$

Proof: Suppose that $Y_{n-2}=y$ and $Z_{n-1}=z>0$. Then

$$\mathbf{E}(\hat{m}_n|Y_{n-2}=y,Z_{n-1}=z)=\mathbf{E}\left(\frac{y+z+Z_n-1}{y+z}\middle|Y_{n-2}=y,Z_{n-1}=z\right)$$

$$=1+\frac{zm-1}{y+z}\,.$$

Averaging over y and $z > 0$ we get

$$\mathbf{E}(\hat{m}_n | Z_{n-1} > 0) = 1 - \mathbf{E}(Y_{n-1}^{-1} | Z_{n-1} > 0) + m\mathbf{E}(Z_{n-1} Y_{n-1}^{-1} | Z_{n-1} > 0) .$$

□

Corollary: If $m \leq 1$

$$\mathbf{E}(\hat{m}_n | Z_n > 0) > m .$$

Proof: Let $m < 1$. From the proposition we have that

$$\mathbf{E}(\hat{m}_n) - m = m\left(\mathbf{E}\left(\frac{Z_{n-1}}{Y_{n-1}}\bigg| Z_{n-1} > 0\right) - 1\right) - \left(\mathbf{E}\left(\frac{1}{Y_{n-1}}\bigg| Z_{n-1} > 0\right) - 1\right)$$

$$\geq (m - 1)\left[\mathbf{E}\left(\frac{1}{Y_{n-1}}\bigg| Z_{n-1} > 0\right) - 1\right] > 0 .$$

The case $m = 1$ is similar. □

The same argument as in Proposition 2.3 will yield the unconditional mean and variance of $\hat{m}_n$. Unconditionally we shall see that $\hat{m}$ underestimates m on average.

Proposition 2.4

(i) $$\mathbf{E}\hat{m}_n = 1 + m\mathbf{E}\,\frac{Z_{n-1}}{Y_{n-1}} - \mathbf{E}\,\frac{1}{Y_{n-1}} .$$

Furthermore, for $n \geq 2$

(ii) $$m - h(m, \tau, n) < \mathbf{E}\hat{m}_n < m ,$$

where

$$h(m, \tau, n) = -\sum_{i=1}^{n-1} [-\tau]^i \frac{a_i(m, n)}{a_0(m, n-1)} ,$$

where $\tau = \sigma^2 / z_0$ and the coefficients $a_i(m, n)$ are given by the recursion formula

$$a_i(m, n) = m a_i(m, n-1) + a_0(m, n-1) a_{i-1}(m, n-1) ,$$
$$1 \leq i \leq n-1 ,$$

with $a_{n-1}(m, n-1) = 0$, $a_0(m, n-1) = \Sigma_{k=0}^{n-1} m^k$ and $a_1(m, 2) = 1 + m$. In addition

(iii) $$\mathbf{Var}\,\hat{m}_n = \sigma^2 \mathbf{E}(Y_{n-1}^{-1}) + m^2\,\mathbf{Var}(Y_{n-1}^{-1}) .$$

Proof: We need only verify the inequalities in (ii). By repeated conditioning we get

$$\mathbf{E}\left(\frac{Z_1+\cdots+Z_n}{Z_0+\cdots+Z_{n-1}}\right)$$

$$=\mathbf{E}\left(\cdots\mathbf{E}\left(\mathbf{E}\left(\frac{Z_1+\cdots+Z_n}{Z_0+\cdots+Z_{n-1}}\bigg|Z_0,\ldots,Z_{n-1}\right)\bigg|Z_0,\ldots,Z_{n-2}\right)\cdots\bigg|Z_0\right).$$

Furthermore

$$\mathbf{E}\left(\frac{Z_1+\cdots+Z_n}{Z_0+\cdots+Z_{n-1}}\bigg|Z_0,\ldots,Z_{n-1}\right)$$

$$=\frac{Z_1+\cdots+Z_{n+1}}{Z_0+\cdots+Z_{n-1}}+\frac{\mathbf{E}(Z_n|Z_0,\ldots,Z_{n-1})}{Z_0+\cdots+Z_{n-1}}$$

$$=\frac{Z_1+\cdots+Z_{n-2}+Z_{n-1}(1+m)}{Z_0+\cdots+Z_{n-1}}. \tag{1}$$

The right-hand side of this expression is a strictly convex function in Z_{n-1}, so by Jensen's inequality

$$\mathbf{E}\left(\frac{Z_1+\cdots+Z_{n-2}+Z_{n-1}(1+m)}{Z_0+\cdots+Z_{n-1}}\bigg|Z_0,\ldots,Z_{n-2}\right)$$

$$<\frac{Z_1+\cdots+Z_{n-3}+Z_{n-2}(1+m+m^2)}{Z_0+\cdots+Z_{n-3}+Z_{n-2}(1+m)}.$$

Repeated application of this technique yields

$$\mathbf{E}\hat{m}_n<\frac{mZ_0(1+m+\cdots+m^{n-1})}{Z_0(1+m+\cdots+m^{n-1})}=m.$$

To get the lower bound we use the following inequality for a nondegenerate positive random variable X and positive constants a, b, c, d (Exercise 4).

$$\mathbf{E}\left(\frac{a+bX}{c+dX}\right)>\frac{a+b\mathbf{E}X}{c+d\mathbf{E}X}-\frac{bd\,\mathbf{Var}\,X}{(c+d\mathbf{E}X)^2}. \tag{2}$$

Starting from (1) we get (for $n\geq 3$)

$$\mathbf{E}\left(\frac{Z_1+\cdots+Z_{n-2}+Z_{n-1}(1+m)}{Z_0+\cdots+Z_{n-1}}\bigg|Z_0,\ldots,Z_{n-2}\right)$$

$$>\frac{Z_1+\cdots+Z_{n-2}(1+m+m^2)}{Z_0+\cdots+Z_{n-3}+Z_{n-2}(1+m)}-\frac{(1+m)\sigma^2Z_{n-2}}{(Z_0+\cdots+Z_{n-2}(1+m))^2}.$$

The second term on the right-hand side can be estimated below, again using convexity, by

$$-\frac{(1+m)\tau Z_{n-2}}{Z_0+\cdots+Z_{n-2}(1+m)}.$$

We continue this reduction for Z_{n-2}, and the result follows by induction. □

Remark: Eschenbach and Winkler (1975) give a similar inequality where

$$h(m,\tau,n)=\tau a_1(m,n)=\tau\sum_{k=1}^{n-1} m^{n-1-k}\left(\sum_{i=0}^{k-1} m^i\right)\left(\sum_{j=0}^{k} m^j\right).$$

The band in Proposition 2.4(ii) is tighter. The inequalities are only useful for small values of n and τ. In Figure 2.2 we illustrate the bounds of the Proposition for $n=3$ and $\tau=0.1$ or 0.5 □

The estimator $\hat{m}_n$ has a lot of intuitive appeal: we have observed Y_{n-1} iid random variables, with a sum of Y_n-1, so we just compute the average

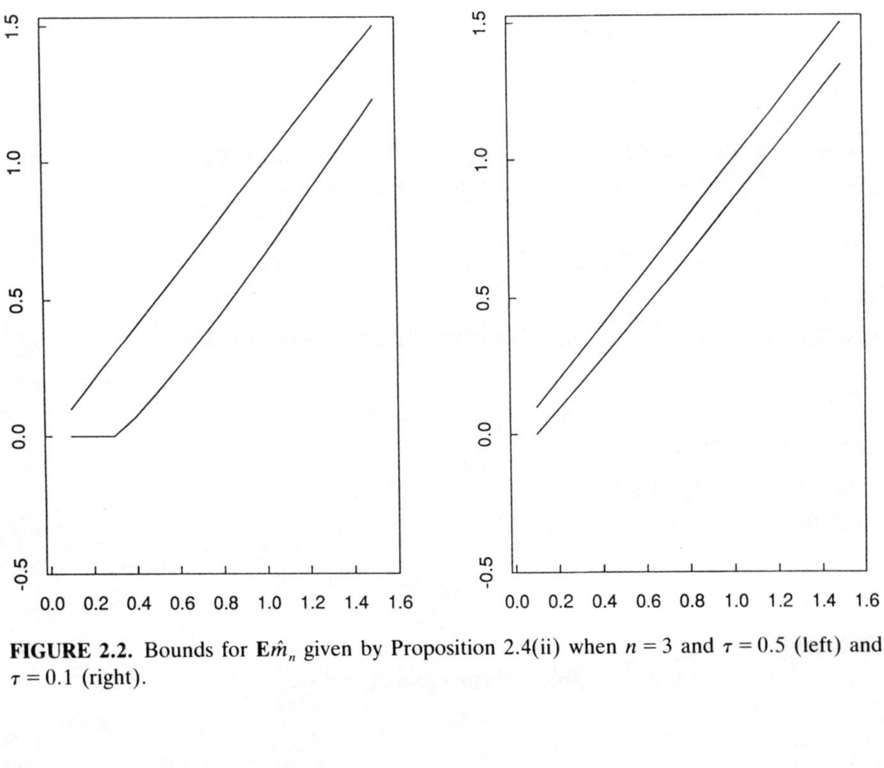

FIGURE 2.2. Bounds for $\mathbf{E}\hat{m}_n$ given by Proposition 2.4(ii) when $n=3$ and $\tau=0.5$ (left) and $\tau=0.1$ (right).

number of offspring per parent. There are other ways of deriving this estimator, as we will see in the next two sections.

3. CONDITIONAL LEAST SQUARES

The method of least squares generalizes in a relatively straightforward way to dependent data. Suppose we have a model of the form

$$X_k = f_k(\theta) + e_k ,$$

where the f_k are $\mathcal{F}_{k-1}$-measurable functions, $\mathcal{F}_k$ is a history of the x_k and the e_k are noise terms, satisfying $\mathbf{E}(e_k|\mathcal{F}_{k-1}) = 0$ and $\mathbf{Var}(e_k|\mathcal{F}_{k-1}) = \nu_k < \infty$. The optimal $\mathcal{F}_{k-1}$-measurable predictor of X_k is seen to be $f_k(\theta)$, if θ is known. If θ is unknown, but we have a sample $X_1, \ldots, X_n$ of observations, we estimate θ by minimizing the weighted prediction error

$$S_n(\theta) = \sum_{k=1}^{n} \nu_k^{-1}(X_k - f_k(\theta))^2 .$$

General theory of the weighted conditional least squares methods can be found in Klimko and Nelson (1978). However, their theory is developed for the case of ergodic processes, and cannot be applied to our situation.

In order to use the least squares method to estimate the offspring mean in a BGW process, notice that

$$Z_n = mZ_{n-1} + e_n , \tag{3}$$

where e_n is the prediction error $Z_n - mZ_{n-1}$. We have $\mathbf{E}(e_n|\mathcal{F}_{n-1}) = 0$ and $\mathbf{Var}(e_n|\mathcal{F}_{n-1}) = \sigma^2 Z_{n-1}$. Thus we want to minimize

$$S_n(m) = \sum_{k=1}^{n} Z_{k-1}^{-1}(Z_k - mZ_{k-1})^2 .$$

It follows that the least squares estimator solves the equation

$$\sum_{k=1}^{n} (Z_k - mZ_{k-1}) = 0 ,$$

which again yields the estimator $\hat{m}_n$.

A different way of writing (3) is

$$Z_n = mZ_{n-1} + \sqrt{Z_{n-1}}\, \epsilon_n .$$

Here $\epsilon_n = e_n/\sqrt{Z_{n-1}}$, and is defined to be 0 if $Z_{n-1} = 0$. If the ϵ_n are uncorrelated random variables, with ϵ_n uncorrelated with Z_{n-1}, this representation is similar to an autoregressive process, except that the innovations ϵ_n are weighted. Following a large observation the process tends to move in large steps, while following a small one it tends to stay put. This type of process was introduced by Engle (1982), and is called a *conditionally heteroscedastic autoregression* (CHAR). These processes are capable of describing patchy behavior in a time series, and have been found useful to describe some economic time series. Writing

$$\mathbf{Cov}(\epsilon_j, Z_{j-1}) = \mathbf{E}(\mathbf{Cov}(\epsilon_j, Z_{j-1}|Z_{j-1})) + \mathbf{Cov}(\mathbf{E}(\epsilon_j|Z_{j-1}), Z_{j-1})$$

and noting that $\mathbf{E}(\epsilon_j|Z_{j-1}) = 0$, we see that ϵ_j is uncorrelated with Z_{j-1}. It follows (using the Markov property) that $\mathbf{Cov}(\epsilon_j, \epsilon_{j-k}) = 0$, $k = 1, \ldots, j-1$. Thus the weighted residuals (ϵ_k) do form an innovations sequence, and the BGW proceess can be thought of as a CHAR process of order one.

Most of the theoretical work on CHAR processes has been concerned with conditions under which these processes are in the stationary domain. It follows from this work (e.g., Nemec, 1984) that the particular form of heteroscedasticity of the innovations that yields a BGW process is of limited general interest, since any solution in the stationary domain (here the case $m < 1$) must converge to zero a.s. Not much work has been done on other nonstationary CHAR processes and their statistical inference.

4. MAXIMUM LIKELIHOOD ESTIMATORS

In order to compute a maximum likelihood estimator (mle) of m we need to specify the underlying distribution. To begin with, let us consider the class of all offspring distributions. This is the nonparametric situation. Assume further that we can observe the entire family tree, and in particular the individual X_{ij}, $i = 1, \ldots, Z_{j-1}$, $j = 1, \ldots, n$. Let

$$N_k(n) = \#\{X_{ij} = k, i \leq Z_{j-1}, j \leq n\} .$$

The parameter is $p = (p_0, p_1, \ldots)$, and the likelihood is

$$L(p) = \prod_{k=0}^{\infty} p_k^{N_k(n)} ,$$

subject to the constraints that $\Sigma\, p_k = 1$, $p_k \geq 0$. Using a Lagrange multiplier for the first constraint we get the partial derivatives of the log likelihood $l(p) = \log L(p)$:

$$\frac{\partial}{\partial p_k}\left\{l(p) + \lambda\left(\sum p_i - 1\right)\right\} = \frac{N_k(n)}{p_k} - \lambda ,$$

$$\frac{\partial}{\partial \lambda}\left\{l(p) + \lambda\left(\sum p_i - 1\right)\right\} = \sum p_i - 1 ,$$

with solutions

$$\hat{\lambda} = \sum N_k(n) = Y_{n-1} ,$$

$$\hat{p}_k = N_k(n) / Y_{n-1} .$$

We see that the $\hat{p}_k$ are nonnegative, so the second constraint is automatically satisfied. This, again, is an intuitively very reasonable estimator: we just estimate p_k by the relative proportion of parents with k offspring. Using the invariance of maximum likelihood estimators under reparametrization we see that

$$\hat{m} = \sum k\hat{p}_k = (Y_n - 1) / Y_{n-1} .$$

Thus $\hat{m}$ is the nonparametric maximum likelihood estimator based on the entire family tree up to the nth generation. Since it is only a function of generation sizes, one would perhaps think that $\hat{m}$ would also be the maximum likelihood estimator based on observing only generation sizes. This is not an automatic conclusion, as the following example shows.

Example: Estimating the Normal Variance

Let $X_1, \ldots, X_n$ be iid $\mathbf{N}(\mu, \sigma^2)$. The mle of σ^2 is $\Sigma (X_i - \bar{X})^2/n$. However, if we only observe $\Sigma (X_i - \bar{X})^2$, this random variable is distributed as $\sigma^2\chi^2_{n-1}$, so the mle is $\Sigma (X_i - \bar{X})^2/(n-1)$. □

It so happens that $\hat{m}$ is indeed an mle when observing only generation sizes. Two different approaches have been used to prove this. Feigin (1977) used a Lagrange multiplier approach, which is outlined in Exercise 7. Keiding and Lauritzen (1978) prove a general theorem, which when applied to the present situation delineates a wide class of models in which $\hat{m}$ is the mle. The most interesting class is the nonparametric one, where we allow any distribution on the nonnegative integers. For a while, we will deviate from the branching process setup, and talk about the iid situation.

Proposition 2.5: Let $Y_1, \ldots, Y_n$ be iid on the nonnegative integers. Then the nonparametric maximum likelihood estimator of $\mathbf{E}Y_1$ is $\bar{Y}$, whether we observe the individual random variables or only their sum.

Proof: We will start by looking at the problem of estimation of the mean of a discrete distribution, supported on $\{0, \ldots, k\}$. Suppose that y is an observation from this distribution. Observing y is equivalent to observing

$$(1(y = 0), \ldots, 1(y = k)) \in X = \left\{(x_0, \ldots, x_k);\, x_i = 0 \text{ or } 1, \sum x_i = 1\right\} .$$

Let $p = (p_0, \ldots, p_k)$ be a vector of probabilities, and let

$$S_p = \{i;\ p_i > 0\}\ .$$

Then S_p is a nonempty subset of the integers $\{0, \ldots, k\}$. Let the lower and upper limit of S_p be l_p and u_p, respectively. Define $t(\mathbf{x}) = \Sigma\, ix_i$, so t is the actual observation. We now want to write the general distribution on $\{0, \ldots, k\}$ in something akin to exponential form. In order to do so we will over-parametrize it, introducing an extra parameter $\xi \in \mathbf{R}$. The density of a general distribution with respect to counting measure on X can be written

$$\phi(\theta)^{-1} e^{\xi t(\mathbf{x})} \prod_{i \in S_p} p_i^{x_i} \qquad \text{if } t(\mathbf{x}) \in S_p\ , \tag{4}$$

where $\theta = (p, \xi)$. This yields a parametrization of the general distribution on $\{0, \ldots, k\}$ for each fixed ξ. In fact, given a distribution $q_0, \ldots, q_k$ on $\mathbf{X}$, and a value of ξ, the p_i's needed to obtain this distribution are

$$p_j = \frac{q_j e^{-\xi j}}{\sum_k q_k e^{-\xi k}}$$

for $j \in S_q$. Also notice that for a given p

$$\phi(\theta) = \int e^{\xi t(\mathbf{x})} \prod_{i \in S_p} p_i^{x_i} \mu(d\mathbf{x}) < \infty$$

for any real ξ. The model (4) is an example of what Barndorff–Nielsen (1978) calls *partly exponential* models. Notice that for a given p the model (4) *is* an exponential family, with sufficient statistic t. Let $\mathbf{P}_\theta$ denote the distribution on $\{0, \ldots, k\}$ with parameter θ (and density (4)), and let $\Pi = \{\mathbf{P}_\theta;\ \theta \in \Theta\}$, $\Pi_p = \{\mathbf{P}_\theta;\ \theta = (p, \xi),\ \xi \in \mathbf{R}\}$. Since $l_p \leq t(\mathbf{x}) \leq \mu_p$, it follows from exponential family theory that $t(\mathbf{x})$ is the mle of $\tau(\theta) = \mathbf{E}_\theta t(\mathbf{X})$ in Π_p, at least whenever $l_p < t(\mathbf{x}) < \mu_p$ (so that $t(\mathbf{x})$ is in the interior of the convex hull of the support S_p; cf. Barndorff–Nielsen, 1978, Theorem 9.13). Next observe that the marginal density of t is

$$\sum_{\{\mathbf{x};t(\mathbf{x})=t\}} \phi(\theta)^{-1} e^{\xi t(\mathbf{x})} \prod_{i \in S_p} p_i^{x_i} \propto e^{\xi t}\ ,$$

which is proportional to the likelihood of $\mathbf{x}$ in Π_p. Thus whenever t is the mle from observing $\mathbf{x}$, it is also the mle from observing t(x) only. But if t is the mle in the family Π_p, it is also the mle in the full family Π. To see this, first notice that $t(\mathbf{x})$ is the mle of $\tau(\theta)$ if

$$\sup_{\theta \in \Theta} L(\theta) = \sup_{\theta \in \tau^{-1}(t(\mathbf{x}))} L(\theta)\ ,$$

where $L(\theta)$ is the likelihood based on observing $\mathbf{x}$. Now compute

$$\sup_{\theta\in\Theta} L(\theta) = \sup_{p} \sup_{\xi} L(\theta) = \sup_{p} \sup_{\psi^{-1}(p)\cap\tau^{-1}(t(\mathbf{x}))} L(\theta) = \sup_{\theta\in\tau^{-1}(t(\mathbf{x}))} L(\theta),$$

where $\psi(\theta) = p$. This is because in order to maximize the likelihood over all θ we can maximize it first over all ξ and then over all p. For any given p we need only look at the maximum over the ξ corresponding to that particular p, i.e., the mle in Π_p. Notice that we need not assume that θ is identifiable from observing $t(\mathbf{x})$ alone; on other words, $\tau^{-1}(t(\mathbf{x}))$ may be a set.

Next we will see that the mle of $\tau(\theta)$ based on n iid observations $\mathbf{x}^{(1)}, \ldots, \mathbf{x}^{(n)}$ from the distribution (4) is $\Sigma\, t(\mathbf{x}^{(i)})/n$. The likelihood is

$$\phi(\boldsymbol{\theta})^{-n} \prod_{i=1}^{n} \left(\prod_{j\in S_p} p_j^{x_j^{(i)}} e^{\xi t(\mathbf{x}^{(i)})} \right) = g(\mathbf{x}^{(1)}, \ldots, \mathbf{x}^{(n)}; p) e^{\xi\, \Sigma\, t(\mathbf{x}^{(i)})} .$$

This is again of the partly exponential form, so the same argument as in the case $n = 1$ shows that the mle of $\tau(\theta)$ must be $\Sigma\, t(\mathbf{x}^{(i)})/n$, whether based on observing all the $\mathbf{x}^{(i)}$ or based only on observing $\Sigma\, t(\mathbf{x}^{(i)})$.

Also notice that if $\Sigma\, t(\mathbf{x}^{(i)})/n = l_p$, we must have $x_{l_p}^{(i)} = 1$ for all i, whence the mle must put mass 1 at l_p. A similar argument holds for the upper bound.

If we have independent observations $y_1, \ldots, y_n$ of a general distribution on the integers, the likelihood when $\Sigma\, y_i = y$ is

$$L(p_0, p_1, \ldots) = \sum_{\Sigma\, ix_i = y} p_0^{x_0} p_1^{x_1} \cdots ,$$

where $x_i = \Sigma_{j=1}^{n} x_i^{(j)}$. This likelihood does not depend on $p_{y+1}, p_{y+2}, \ldots,$ and is increasing in each of the first $y + 1$ coordinates. Hence the maximization problem is identical to that of the general distribution on $\{0, \ldots, y\}$. □

There are many more classes of distributions for which a result similar to that of Proposition 2.5 holds. The same arguments as in the proof of the proposition yield the following result.

Corollary 1: Suppose that $\theta = (\omega, \xi)$ and that P_θ has density $\phi(\theta)^{-1} g(x, \omega) \exp(\xi t(x))$ with respect to an appropriate dominating measure. Let $\psi(\theta) = \omega$ and $\tau(\theta) = \mathbf{E}_\theta t(X)$. Then, if $t(x)$ is in the interior of the closed convex hull of the support of the distribution of $t(X)$, the maximum likelihood estimator $\hat{\tau}(\theta)$ is $t(x)$ whether x or $t(x)$ has been observed.

The general result given in Corollary 1 applies whenever we have observations from a partial exponential family, such as the general (nonparametric) distribution on the integers, the power series distribution (or one-parameter discrete exponential family), the negative binomial distribution, or the geometric distribution with modified zero term.

Example: The Zeta Distribution

A situation where $\hat{m}$ is not the mle of m is the zeta distribution (or discrete Pareto distribution)

$$p_k(\theta) = (k+1)^{-\theta}/\zeta(\theta) , \qquad \theta > 1 ,$$

where $\zeta(\cdot)$ is the Riemann zeta function. This has found some applications in linguistics to describe word frequencies, where it is also called the Zipf–Estoup law (cf. Good, 1957, for a discussion of this), in ecology to describe species numbers, and in insurance to describe the number of policies held by an individual. The mean is $m(\theta) = \zeta(\theta - 1)/\zeta(\theta) - 1$, and the geometric mean of the $(1 + x_i)$ is a sufficient statistic. Hence an estimator based on $\Sigma\, x_i$ cannot be maximum likelihood. In fact, the likelihood equation for θ based on a sample of size n is

$$-\frac{\zeta'(\theta)}{\zeta(\theta)} = \frac{1}{n} \sum_i \log(1 + x_i) .$$

The loss in efficiency in using $\hat{m}$ rather than $m(\hat{\theta})$ can be substantial for m near 1. □

A slight generalization of Corollary 1 is fairly easy, and will be needed to apply these results to the branching process case.

Corollary 2: The results of Proposition 2.5 and Corollary 1 hold also for random n, provided that n is a stopping rule whose definition does not explicitly depend on θ.

Proof: This result is immediate from the proposition and a well-known (but seldom proved) result, namely that the likelihood is independent of the stopping rule.

Lemma 2.2: Let (ξ_n) be a stochastic process, and define the natural history $\mathscr{F}_n = \sigma\{\xi_0, \ldots, \xi_n\}$. Let N be a stopping time with respect to $\mathscr{F}_n$, so that $\{N = n\} \in \mathscr{F}_n$. Assume that the process distribution $\mathbf{P}_\theta$, dominated by some σ-finite measure μ, depends on the parameter θ, but that the indicator function $1(N = n)$ can be evaluated without knowing θ. Then the maximum likelihood estimator of θ based on $\xi_0, \ldots, \xi_N$ on the set $\{N = n\}$ is the same as the maximum likelihood estimator of θ based on $\xi_0, \ldots, \xi_n$.

Proof of Lemma: Let $f(x_0, \ldots, x_n; \theta)$ be the Radon–Nikodym derivative of the distribution of $\xi_0, \ldots, \xi_n$ with respect to μ. Notice that $\mathbf{P}_\theta\{N = n | \mathscr{F}_n\} = 1(N = n)$. Then

$$\begin{aligned}
&\mathbf{P}_\theta\{\xi_0 \in A_0, \ldots, \xi_n \in A_n, N = n\} \\
&\quad = \mathbf{P}_\theta\{\xi_0 \in A_0, \ldots, \xi_n \in A_n\}\mathbf{P}_\theta\{N = n | \xi_0 \in A_0, \ldots, \xi_n \in A_n\} \\
&\quad = \int_{A_0 \times \cdots \times A_n} f(x_0, \ldots, x_n; \theta) 1(N = n) \mu\,(dx_0, \ldots, dx_n)\,.
\end{aligned}$$

Thus the likelihood of $\xi_0 = x_0, \ldots, \xi_N = x_N$ on $\{N = n\}$ (i.e., the Radon–Nikodym derivative) is $f(x_0, \ldots, x_n; \theta)1(N = n)$, which is proportional to the likelihood $f(x_0, \ldots, x_n; \theta)$ based on a fixed length sample path. □

Now let us look at the implication of these results for the BGW estimation problem. Suppose that the offspring distribution is such that the mle of m based on observing $X_1 + \cdots + X_k$ is $\bar{X}$. Then $\hat{m}$ is the mle of m based on observing only $Z_1, \ldots, Z_n$. This follows immediately from Corollary 2 and the representation of a BGW process using a random walk (see Exercise 1.6). Also notice that $\bar{m}_n$ is the mle based on observing only Z_{n-1} and Z_n. Thus, intuitively at least, $\hat{m}$ uses more information than $\bar{m}$. We will later verify this in a more formal way.

Example: The Modified Geometric Distribution

Suppose that we observe the complete family tree from a modified geometric distribution having

$$\mathbf{P}(X = k) = p_0 1(k = 0) + (1 - p_0)(1 - c)c^{k-1}1(k > 0)\,.$$

It is easy to see that the likelihood becomes

$$L(p_0, c) = p_0^{N_0(n)}[(1 - p_0)(1 - c)]^{Y_{n-1} - N_0(n)} c^{Y_n - Z_0}\,,$$

where $N_0(n)$ is the total number of individuals with zero offspring. The mle of p_0 is $\hat{p}_0 = N_0(n)/Y_{n-1}$, and the mle of c is $\hat{c} = (Y_n - Z_0)/(Y_n + Y_{n-1} - Z_0 - N_0(n))$. Consequently,

$$\hat{m} = (1 - \hat{p}_0)/(1 - \hat{c}) = (Y_n - Z_0)/Y_{n-1}\,.$$

From the results above we conclude that this is also the mle of m based on observing only generation sizes. □

Application: The Whooping Crane Population of North America

The whooping crane is a very rare migratory bird with breeding grounds in Canada's Northwest Territories and wintering grounds in Texas. Miller et al. (1974) give annual counts from 1938–1972 of whooping cranes arriving in Texas in the fall. Figure 2.3 shows the data. Using a linear birth and death process to describe the population is equivalent to studying a branching process with modified geometric offspring distribution. In terms of the parameters of the underlying distribution we have

$$p_0 = \frac{\mu - \lambda}{\mu - \lambda\tau}, \qquad c = \frac{(\mu - \lambda)\tau}{\mu - \lambda\tau},$$

where $\tau = \exp(\lambda - \mu)$. Since the data are total counts, we do not observe the sufficient statistic $(N_0(n), Y_{n-1}, Y_n)$. The contributions to the likelihood from each generation therefore will be an average over the possible values of the number of zero offspring individuals in the previous generation. It is not hard to verify that the transition probability from $Z_i = k$ to $Z_{i+1} = j$ is

$$p(j|k) = \sum_{i=\max(0,k-j)}^{k} \binom{k}{i} \binom{j-1}{k-i-1} p_0^i[(1-p_0)(1-c)]^{k-i} c^{j-k+i}.$$

Numerically maximizing the likelihood we get $\hat{\lambda} = 2.65$ and $\hat{\mu} = 2.56$ corresponding to $\hat{p}_0 = 0.265$ and $\hat{m} = 1.035$.

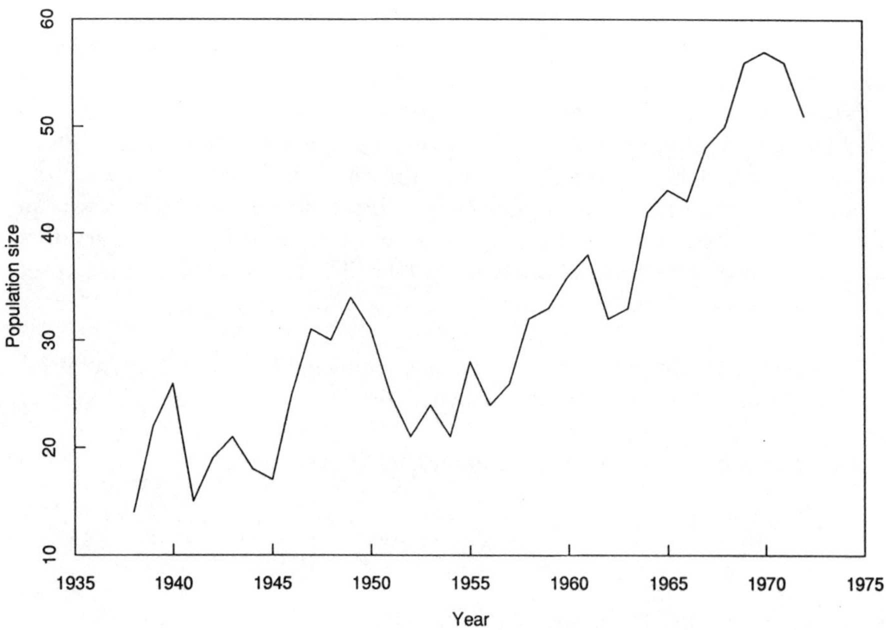

FIGURE 2.3. Annual counts of whooping cranes arriving in Texas 1938–1972.

It is possible to get a rough idea of N_0, since the data contain counts of young birds, who have a different plumage, and adult birds. This gives an estimate of the number of adult birds that die between seasons. Assuming that those birds had no offspring, that the young birds had no offspring, and that all other birds had offspring, we can estimate N_0. The resulting estimate of p_0 is 0.262, which is closer to the mle than we have any right to expect. It should be pointed out that the counts certainly are not correct, as the computed number of deaths for the years 1939 and 1946 are negative. The discussion in Keiding (1976) suggests that an immigration-death process would be more suitable for this population, and Exercise 5.3 is concerned with estimation in that model. □

5. CONSISTENCY

From Theorem 1.1 we know that there are no consistent estimators. For example, whenever $m \leq 1$ we have that $\hat{m}_n \to 1 - 1/Y_\infty$, where Y_∞ is the total size of the population. Also, $\bar{m}_n \to 1$ for $m < 1$, since the generation sizes will be zero a.s. from some time on. These facts are also true for $m > 1$ on the extinction set. In much of iid estimation theory, the idea of consistency is taken as a weak necessary requirement of a sequence of estimators. When the data are dependent, the situation is not as clear. In many situations one would want to compute an estimate based on a single realization of the process (we will discuss the simpler case of inference from several independent realizations later). The notion of *ergodic* sequences was introduced in statistical physics by Boltzmann to denote processes for which time averages were consistent estimators of the population average (the averages from independent realizations of the same proces are called *ensemble* averages). Technically, this amounts to requiring that the dependence between observations far apart is very small. In turn, this implies a zero-one law for tail events. Since the tail σ-field of a BGW process is nontrivial, the best we can hope for is consistency on some tail sets. We will show that the estimators $\hat{m}_n$ and $\bar{m}_n$ are sometimes consistent on $\{W > 0\}$, i.e., on the explosion set A.

Theorem 2.1: Suppose $m > 1$. Then $\hat{m}_n$ and $\bar{m}_n$ both are consistent on A. In addition $\bar{m}_n$ is consistent also when $m = 1$.

Proof: For $\hat{m}_n$ we use the Corollary to Theorem 1.1:

$$\hat{m}_n = \frac{Y_n - 1}{Y_{n-1}} = m \frac{m^{-n}(Y_n - 1)}{m^{-(n-1)}Y_{n-1}} \to m \frac{mW/(m-1)}{mW/(m-1)} = m$$

since W is nonzero on A a.s. by Theorem 1.1(iii). For $\bar{m}_n$ use Theorem 1.1(ii and iii) much like in the argument for $\hat{m}_n$. The case $m = 1$ follows since we defined $\bar{m}_n$ to be 1 whenever Z_{n-1} is 0, which will eventually happen. □

By Theorem 1.1 we know that A is equivalent to the set $\{W > 0\}$. A more interesting equivalence is the following.

Lemma 2.3: Let (B_n) be a sequence of $\mathcal{F}_n$-measurable sets. Then

$$\lim_{n\to\infty} |\mathbf{P}(B_n|A) - \mathbf{P}(B_n|Z_n > 0)| = 0 .$$

Proof

$$\mathbf{P}(B_n|Z_n > 0) = \frac{\mathbf{P}(B_n, A, Z_n > 0)}{\mathbf{P}(Z_n > 0)} + \frac{\mathbf{P}(B_n, A^c, Z_n > 0)}{\mathbf{P}(Z_n > 0)} ,$$

so

$$\begin{aligned}
&\mathbf{P}(B_n|A) - \mathbf{P}(B_n|Z_n > 0) \\
&\quad = \frac{\mathbf{P}(Z_n > 0)(\mathbf{P}(B_n, A) - \mathbf{P}(B_n, Z_n > 0, A))}{\mathbf{P}(A)\mathbf{P}(Z_n > 0)} \\
&\qquad + \frac{\mathbf{P}(B_n, Z_n > 0, A)(\mathbf{P}(Z_n > 0) - \mathbf{P}(A))}{\mathbf{P}(A)\mathbf{P}(Z_n > 0)} - \frac{\mathbf{P}(B_n, Z_n > 0, A^c)}{\mathbf{P}(Z_n > 0)} .
\end{aligned}$$

Since $\{Z_n > 0\} \downarrow A$, we get

$$|\mathbf{P}(B_n, Z_n > 0, A) - \mathbf{P}(B_n, A)| \to 0$$

and

$$\mathbf{P}(B_n, Z_n > 0, A^c) \to 0 .$$

□

Lemma 2.3 and Theorem 2.1 yield that we have consistency under the condition $Z_n > 0$. This is (in a sense) a *verifiable* condition, as opposed to the somewhat nebulous condition A. However, the event $\{Z_n > 0\}$ has probability that depends on the unknown underlying distribution, and in particular depends on the value of the parameter m that we are trying to estimate. It is often considered improper to condition on random variables whose distribution depends directly on the unknown parameter, though there are some definitions of ancillarity that allow a limited amount of such dependence (see Barndorff–Nielsen, 1978, and the discussion in section 9). The crucial feature of the kind of statistical models where you do not necessarily get infinite amounts of information seems to be that inference may be conditional on tail events. Lemma 2.3 gives us a finite sample equivalence that allows us to specify exactly which tail event we are discussing. This is related to the concept of *relevant subsets* in interval estimation: conditioning on a relevant subset changes the distribution of the pivotal quantity (cf. Fisher, 1973). In our situation, $\{Z_n > 0\}$ is a relevant subset of the sample space.

When $m < 1$ the set A has probability 0. However, the following result shows that limits conditional on $Z_n > 0$ still can give interesting results.

Proposition 2.6: Let $m < 1$. Then

$$\lim_{n\to\infty} \mathbf{P}(Z_n = j | Z_n > 0) = p_j^* ,$$

where $\Sigma\, s^k p_k^* = Q(s)$ is a pgf satisfying

$$Q(P(s)) = mQ(s) + 1 - m .$$

Proof

$$Q_n(s) = \mathbf{E}(s^{Z_n}|Z_n > 0) = 1 - \frac{1 - P_n(s)}{1 - P_n(0)} = 1 - G_n(s)$$

say. If $h(s) = (1 - P(s))/(1 - s)$ we see that

$$G_n(s) = \frac{h(P_{n-1}(s))}{h(P_{n-1}(0))}\, G_{n-1}(s) .$$

Since P_n and h are increasing functions, G_n is also increasing and converges, so $Q_n(s) \to Q(s)$. Notice that $G_n(P(s)) = G_{n+1}(s)h(P_n(0))$, and that $h(P_n(0)) \uparrow m$, so $G(P(s)) = mG(s)$ where $G(s) = \lim G_n(s)$. But $Q(s) = 1 - G(s)$. It remains to show that Q corresponds to a probability measure. Letting $s \to 1$ we see that $G(1-) = mG(1-)$, so since $m < 1$ we have $G(1-) = 0$, i.e., $Q(1-) = 1$. □

From Markov chain theory (e.g., Feller, 1968, chapter XV) we know that a positive recurrent chain has a stationary distribution π, given by the solution to $\pi = \pi P$ where P is the transition matrix and π is a probability distribution. If $\Sigma\, \pi_i = \infty$, π is called a stationary measure. In the branching process case, the only nonzero solution is $\pi = (1, 0, 0, \ldots)$: if $p_0 > 0$ we must have $P_{j0} > 0$ for all j, so

$$\pi_0 = \pi_0 P_{00} + \pi_1 P_{10} + \cdots = \pi_0 + \pi_1 P_{10} + \cdots$$

so $\pi_1 = \pi_2 = \cdots = 0$. If $p_0 = 0$, $P_{ij} = 0$ for $j < i$, so

$$\pi_i = \pi_1 P_{1i} + \cdots + \pi_i P_{ii} , \qquad i = 1, 2, \ldots .$$

Let π_n be the first $\pi_i \neq 0$. Then $\pi_n = \pi_n P_{nn}$, but $P_{nn} = p_1^n < 1$, so $\pi_n = 0$, a contradiction. Hence $\pi = (0, 0, 0, \ldots)$ is the only solution in this case. One can think of the stationary measure as describing the average behavior of an ensemble of independent particles, all moving independently according to the same transition matrix. If we start a random number N_j in state j, where $N_j \sim \text{Po}(\pi_j)$, then the expected number in state k at any time is π_k. In our case this simply means that either $p_0 = 0$, in which case all particles move off to infinity, or $p_0 > 0$, in which case an infinite number will be absorbed at 0. However, we may still have a stationary measure on $1, 2, \ldots$. In fact, provided $0 < p_0 < 1$ there is such a measure, satisfying $\pi_i = \sum_1^\infty \pi_j P_{ji}$, $i = 1, 2, \ldots$ and $\sum \pi_i = \infty$. Its generating function $\Pi(s) = \sum \pi_i s^i$ satisfies Abel's equation

$$\Pi(P(s)) = 1 + \Pi(s)\,, \qquad |s| < q$$

when normalized so that $\Pi(P(0)) = 1$. The proof is similar to that of Proposition 2.6.

Example: Fractional Linear Case

When $P(s) = 1 - b/(1-c) + bs/(1-cs)$ is a fractional linear pgf, a stationary measure on $1, 2, \ldots$ is

$$\pi_i = \frac{1-c}{c}\,, \qquad m = 1$$

$$\pi_i = \frac{1}{\log 1/m}\,\frac{1 - s_0^{-i}}{i}\,, \qquad m < 1$$

and

$$\pi_i = \frac{1}{\log m}\,\frac{s_0^{-i} - 1}{i}\,, \qquad m > 1\,,$$

where s_0 is the solution to $P(s) = s$ not equal to 1. □

If we assume the existence of just over two finite moments, then if $m = 1$ one can show that $\lim \pi_i = 2/\sigma^2$ (Kesten et al., 1966). Assuming four finite moments, Karlin and McGregor (1967) proved more precisely that

$$\pi_i = \frac{2}{\sigma^2} - \left(1 - \frac{2P^{(3)}(1)}{3\sigma^4}\right)\frac{1}{i} + \frac{e_i}{i}$$

where $\sum e_i^2 < \infty$. For the case $m \neq 1$ a weaker result is known (Lipow, 1971): $\sum_1^n \pi_i q^i = \log n/\log(1/\gamma) + o(\log n)$, where $\gamma = P'(q)$.

Application: Random Gene Extinction

An important topic of study in genetics is the relationship between the genetic variability present in a species and the rate of mutations. In order to establish such relations one must first look at the gradual reduction of variability due to random extinction of genes. Suppose we consider a multitude of genetic factors, α, β, $\gamma, \ldots$, etc. For simplicity we will look at factors that have two alleles (or gene types). Denote the alleles for the factor α by a and A, respectively. This implies that the factor has three genotypes: aa, aA, and AA. The alleles a, b, c, ... are called small alleles. We will assume that the population is in equilibrium with N individuals at each generation, and that mating is random. Let x_n be the number of small alleles at factor α in the population at the nth generation. An individual with the aa combination contributes 2 to x_n, etc. A factor becomes fixed when one of the alleles becomes extinct, i.e., in the case of α, when $x_n = 0$ or $2N$. Assume that (x_k) is a Markov chain with transition probabilities $Q_{ij} = \mathbf{P}(x_k = j | x_{k-1} = i)$, and that the same transition matrix applies to the other factors. If we call the expected number of factors in generation n represented by i small alleles $\pi_i(n)$ we have the relation (e.g., Moran, 1962, chapter IV)

$$\pi_i(n) = \sum_{j=1}^{2N-1} Q_{ji} \pi_j(n) .$$

If i is very small (or very large) we note that the most probable mating involving the allele a is aA $\times$ AA (or aa $\times$ aA). Roughly speaking, the number of small allele offspring should be independent variates with the same distribution, and so Q_{ij} should be close to the transition matrix P_{ij} for a branching process. In order to keep a stationary population size we set $m = 1$. As n gets large, $\pi(n) = (\pi_1(n), \pi_2(n) \ldots)$ converges to an eigenvalue of Q. Since Q is close to a branching process, π, describing the proportion of small alleles in each state, should also be close to a stationary distribution for the branching process. By the Kesten et al. (1966) result given above we have that

$$\pi_i(n) \approx \frac{2}{\sigma^2}$$

at least for small n where the branching process approximation is reasonable. Thus the proportion of factors that become fixed in each generation is about $\sigma^2/2N$. If it is reasonable to assume a Poisson offspring distribution, we see that about one factor in every two generations becomes fixed. In the next section we will see that a favorable mutation has a much stronger influence on the genetic structure of a population. □

We can use Proposition 2.6 to establish a property of $\bar{m}_n$ where $m < 1$.

Corollary: If $m < 1$, $\mathbf{Var}(\bar{m}_n | Z_{n-1} > 0) \rightarrow \sigma^2 \, \mathbf{Var}(1/X)$, where X has the distribution Q from the previous proposition.

Proof: This follows immediately from Propositions 2.1 and 2.6. □

Example: The Geometric Distribution

Generally the functional equation in Proposition 2.6 is hard to solve, although it usually does determine Q uniquely. Sometimes the solution can be obtained directly. In the geometric case we use that the pgf $P_n(s)$ of Z_n can be written $P_n(s) = \alpha_n + \beta_n s/(1-\gamma_n s)$, as in the Example in the beginning of section 2.2. Therefore the conditional pgf, given that $Z_n > 0$, is $P_n(s)/(1-\alpha_n)$. Letting $n \to \infty$, and assuming that $m < 1$, we see that $Q(s) = (1-m)s/(1-ms)$, i.e., the conditional limiting distribution is a positive geometric distribution with success probability $1 - m = 2 - 1/p$. You may want to check that Q actually does satisfy the functional equation.

As a numerical example, when $p = 0.55$, so that $m = 0.82$, we find that the limiting conditional distribution is positive geometric with mean 5.5. The variance of $1/X$ is 0.14, and the limiting conditional variance of $\bar{m}_n$ is 0.21, corresponding to a standard error of 0.46. □

6. ASYMPTOTIC DISTRIBUTIONS

Yanev (1975) discusses different ways of obtaining asymptotic distributions. One way is to let the initial sample size get large. We used this method in chapter 1. This yields a structure that is equivalent to a large number of iid chains. Thus the asymptotics are akin of classical estimation theory. Suppose that we have a BGW process starting from z ancestors, and let $\hat{m}_n(z) = (Y_n - z)/Y_{n-1}$.

Theorem 2.2: For fixed n as $z \to \infty$

(i) if $m < \infty$, $\hat{m}_n(z) \to m$ a.s. and $\mathbf{E}\hat{m}_n(z) \to m$
(ii) if $\sigma^2 < \infty$, then

$$\left(\frac{z}{\sigma^2} \frac{(m-1)^2}{\sum_0^{2n} m^i - (2n+1)m^n} \right)^{1/2} (\hat{m}_n(z) - m) \xrightarrow{d} \mathbf{N}(0, 1).$$

Proof: (i) Write $Y_n = Y_n(z) = \Sigma_1^z Y_n^i$ of independent random variables Y_n^i, all with the distribution of the total population size at time n, starting from one ancestor. Then

$$\hat{m}_n(z) = \frac{Y_n(z) - z}{Y_{n-1}(z)} = \frac{\sum_1^z Y_n^i - z}{\sum_1^z Y_{n-1}^i} = \frac{\frac{1}{z}\sum_1^z Y_n^i - 1}{\frac{1}{z}\sum_1^z Y_{n-1}^i}.$$

By the law of large numbers, $1/z\,\Sigma_1^z\, Y_n^i \to \mathbf{E}Y_n^1$ a.s. and in L_1. Thus

$$\hat{m}_n(z) \to \frac{\dfrac{m^{n+1}-1}{m-1} - 1}{\dfrac{m^n - 1}{m-1}} = m$$

a.s. and in L_1.

(ii) Provided that $m \neq 1$, using Slutsky's theorem and the argument in (i) we see that $\hat{m}_n(z)$ is asymptotically equivalent to

$$\frac{\dfrac{1}{z}\sum_1^z Y_n^i - 1}{\dfrac{m^n-1}{m-1}}.$$

Since

$$\left(\frac{z}{\mathbf{Var}\, Y_n^i}\right)^{1/2}\left(\frac{1}{z}\sum_1^z (Y_n^i - 1) - \mathbf{E}(Y_n^i - 1)\right) \xrightarrow{d} \mathbf{N}(0, 1)$$

by the central limit theorem, and we know from Exercise 1.2(b) that

$$\mathbf{Var}\, Y_n^i = \frac{\sigma^2}{(m-1)^2}\left(\frac{m^{2n+1}-1}{m-1} - (2n+1)m^n\right)$$

the statement of the theorem follows. The case $m = 1$ is similar. □

If a large initial sample size does not seem reasonable, we must attempt to do asymptotics from a long stretch of data instead. This, of course, is the basis for the consistency results in the previous section. Since the estimators of m are not consistent of the entire sample space, there is no hope for unconditional asymptotic normality. We will now study the problem of the asymptotic distribution of suitably scaled sequences of estimators of m. It turns out that these limiting distributions are not normal as long as we use deterministic normings. The main tool will be a martingale central limit theorem for martingales whose differences are not asymptotically negligible, but rather grow at a geometric rate.

If we are to get useful conditional limit theorems, we need to introduce the concept of *stable* convergence. We say $Y_n \to Y$ (stable) if

$$\mathbf{P}(Y_n < y, E) \to Q_y(E) \qquad \text{as } n \to \infty$$

for any Borel set E, where $Q_y(E) \to \mathbf{P}(E)$ as $y \to \infty$. If also

$$Q_y(E) = \mathbf{P}(Y \leq y)\mathbf{P}(E)\,,$$

the convergence is said to be *mixing*. Most martingale central limit theorems turn out to be stable, and thus allow us to talk about convergence on a subset of the sample space. In the mixing case, the limiting random variable Y is also independent of the condition E.

Usually functional central limit theorems are given on function spaces such as C[0, 1] or D[0, 1]. In order to state the main theorem of this section we need to look at points in $\mathbf{R}^{\mathbb{N}}$, the space of all real sequences. We make this into a complete, nonseparable metric space by using the metric

$$d(x, y) = \sup_{n \geq 0} \rho\left(\sum_{j=0}^{n} x_j, \sum_{j=0}^{n} y_j\right),$$

where $\rho(a, b) = |a - b|/(1 + |a - b|)$ for real numbers a and b. We can extend any probability measure defined on the usual Borel field (generated by cylinder sets) to one on the Borel field $\mathscr{D}$ generated by d.

Let $(S_n, \mathscr{F}_n)_{n\geq 0}$ be a martingale on a probability space $(\Omega, \mathscr{F}, \mathbf{P})$. Write $S_n = \Sigma_{j=1}^{n} U_j$, $s_n^2 = \mathbf{E}S_n^2 < \infty$ and $V_n^2 = \Sigma_1^n \mathbf{E}(U_j^2 | \mathscr{F}_{j-1})$ where $\mathscr{F}_0 = \{\varnothing, \Omega\}$. For $p > 1$ and $a > 0$, define random sequences in $\mathbf{R}^{\mathbb{N}}$ by

$$\xi_n = (\xi_{n0}, \xi_{n1}, \ldots),$$

where

$$\xi_{nj} = \begin{cases} \dfrac{U_{n-j}}{as_n}, & j = 0, 1, \ldots, n-1, \\ 0, & j \geq n. \end{cases}$$

This is similar to the usual triangular array introduced to prove central limit theorems, except that the indexing is backwards. Let

$$\xi = (\eta Y_0, \eta Y_1, \ldots)$$

where $\eta \geq 0$ is independent of $Y_0, Y_1, \ldots$, which in turn are independent with $Y_j \sim \mathbf{N}(0, p^{-j})$. Also let

$$\zeta_n = (\zeta_{n0}, \zeta_{n1}, \ldots) = s_n V_n^{-1} \xi_n$$

and $\zeta = (Y_0, Y_1, \ldots)$. The ζ_n have a random norming instead of a fixed one. We are now ready to state the martingale central limit theorem of Scott (1978).

Proposition 2.7: Suppose that

$$s_n^{-2} V_n^2 \xrightarrow{\mathbf{P}} \eta^2, \qquad \text{with } \mathbf{E}\eta^2 < \infty, \tag{5}$$

$$s_n^{-2} s_{n-r}^2 \to p^{-r}, \qquad r \geq 0, \tag{6}$$

and

$$\mathbf{E}(e^{itU_n/(aV_n)}|\mathcal{F}_{n-1}) \xrightarrow{\mathbf{P}} \exp(-t^2/2) \qquad \text{on } \{\eta > 0\} . \tag{7}$$

Then for every E in $\mathcal{D}$ such that $\mathbf{P}(\xi \in \partial E) = 0$ and every F in $\mathcal{F}$

$$\mathbf{P}(\xi_n \in E, F|\eta > 0) \to \mathbf{P}(\xi \in E)\mathbf{P}(F|\eta > 0) ,$$

and for every E in $\mathcal{D}$ such that $\mathbf{P}(\zeta \in \partial E) = 0$ and every F in $\mathcal{F}$

$$\mathbf{P}(\zeta_n \in E, F|\eta > 0) \to \mathbf{P}(\zeta \in E)\mathbf{P}(F|\eta > 0) .$$

The difference between this central limit theorem and more usual ones is that here the individual summands are not asymptotically negligible. Rather, the last term dominates all other terms. But this last term is itself a sum of a large number of independent random variables, and hence may be expected to be asymptotically normal. Notice, however, that when nonrandom norming is used (i.e., we look at the sequence ξ_n), the resulting limit distribution is *not* normal, but rather a mixed normal with the distribution of η as mixing distribution. On the other hand, if the random norming V_n is used, the limiting distribution is normal. Later we will interpret this phenomenon in an estimation context.

The application to supercritical BGW processes is relatively straightforward. Let $U_k = Z_k - mZ_{k-1}$. Then $S_n = \Sigma_1^n U_k$ is a martingale with respect to the natural history $\mathcal{F}_n$, with $s_n^2 = \sigma^2(m^n - 1)/(m-1)$, and

$$V_n^2 = \sum_{k=0}^{n} \mathbf{E}(U_k^2|\mathcal{F}_{k-1}) = \sigma^2 Y_{n-1} .$$

Theorem 1.1 and the Toeplitz lemma show that $s_n^{-2}V_n^2 \to W$ a.s., and hence that 5 is satisfied. To verify (6), compute $s_n^{-2}s_{n-r}^2 \to m^{-r}$ for $r \geq 0$. Let $\psi(t) = \exp(-itm)P(it/\sigma)$ be the characteristic function of $(Z_1 - mz)/\sigma$. Since $(Z_n - mZ_{n-1})/\sigma$ is a sum of Z_{n-1} independent random variables with characteristic function ψ, we get

$$\mathbf{E}(\exp(it(Z_n - mZ_{n-1})/Z_{n-1}^{1/2}\sigma)|\mathcal{F}_{n-1}) = (\psi(tZ_{n-1}^{-1/2}))^{Z_{n-1}} .$$

On $\{W > 0\}$ we have that $Z_n \xrightarrow{\text{a.s.}} \infty$, so by the central limit theorem

$$(\psi(tZ_{n-1}^{-1/2}))^{Z_{n-1}} \xrightarrow{\text{a.s.}} \exp(-t^2/2) \qquad \text{on } \{W > 0\} .$$

From Theorem 1.1 and its corollary it follows that

$$Z_{n-1}^{-1}Y_{n-1} \xrightarrow{\text{a.s.}} \frac{m}{m-1} \qquad \text{on } \{W > 0\} ,$$

so that on $\{W>0\}$

$$\mathbf{E}(\exp(it(Z_n - mZ_{n-1})/(m^{-1/2}(m-1)^{1/2}V_n))|\mathscr{F}_{n-1}) \overset{\text{a.s.}}{\to} \exp(-t^2/2)\,,$$

verifying equation (7). Thus if we put

$$\xi_{nj} = \frac{Z_{n-j} - mZ_{n-j-1}}{m^{-1/2}(m-1)^{1/2}s_n}\,, \qquad j<n$$

and

$$\zeta_{nj} = s_n \xi_{nj}/V_n\,,$$

the conditions of Proposition 2.7 are satisfied.

Theorem 2.3: Assume that $m>1$. Let W be as in Theorem 1.1, and let Y be a standard normal random variable, independent of W. Then for any x and any $F \in \mathscr{F}$

$$\mathbf{P}\left(\frac{m^{(n-1)/2}}{\sigma}(\bar{m}_n - m) \le x,\, F|Z_n>0\right) \to \mathbf{P}(Y \le xW^{1/2})\mathbf{P}(F|W>0)$$

and

$$\mathbf{P}\left(\frac{m^{(n-1)/2}}{\sigma(m-1)^{1/2}}(\hat{m}_n - m) \le x,\, F|Z_n>0\right) \to \mathbf{P}(Y \le xW^{1/2})\mathbf{P}(F|W>0)\,.$$

For any x and any $F \in \mathscr{F}$

$$\mathbf{P}\left(\frac{Z_{n-1}^{1/2}}{\sigma}(\bar{m}_n - m) \le x,\, F|Z_n>0\right) \to \mathbf{P}(Y \le x)\mathbf{P}(F|W>0)$$

and

$$\mathbf{P}\left(\frac{Y_{n-1}^{1/2}}{\sigma}(\hat{m}_n - m) \le x,\, F|Z_n>0\right) \to \mathbf{P}(Y \le x)\mathbf{P}(F|W>0)\,.$$

Proof: By Lemma 2.3, the condition $\{Z_n>0\}$ is asymptotically equivalent to $\{W>0\}$. To get the results for $\bar{m}$, write

$$\bar{m}_n - m = \frac{as_n}{Z_{n-1}}\,\xi_{n0}$$

and use the continuous mapping $h:\mathbf{R}^{\mathbb{N}} \to \mathbf{R}$ given by $h(x) = x_0$, together with the continuous mapping theorem (Billingsley, 1968, Theorem 5.1) and Proposition 2.7. To get those for $\hat{m}$, use the mapping $h(x) = \limsup \Sigma_0^{n-1} x_j$, which is continuous by our definition of the metric, and write

$$\hat{m}_n - m = \frac{as_n}{Y_{n-1}} \sum_{i=0}^{n-1} \xi_{ni}\,. \qquad \square$$

Remark: Careful analysis of the proof of Proposition 2.7 reveals that the result of Theorem 2.3 is uniform in $m \geq 1 + \epsilon$ for any $\epsilon > 0$ (Heyde, 1978). □

Application: Smallpox Epidemics

In 1956 there was an outbreak of the less serious form of smallpox, *variola minor*, in Vila Guarani, a semi-isolated residential worker's district of São Paulo in Brazil. The disease was imported by two visitors, who soon after left the community. Smallpox can be modeled fairly well by a BGW process (Becker, 1972; see also the discussion in section 1). There is a fairly long latent period, 2–16 days, between infection and the onset of a rash. An infective is infectious from the time of appearance of the first lesions until all scabs have disappeared, a duration of about 3 weeks. *Variola minor* is lethal in about 10% of the cases. The times of reported symptoms in Vila Guarani show a clear clustering, due to the latent period of the disease. Since the disease was introduced by visitors, whose contact pattern with the rest of the population may be quite different from that of permanent residents, we take the ancestor to be the first infected resident. The first five cluster sizes are 1, 5, 3, 12, and 24. After the fifth cluster, which we will take as the fourth generation, mass vaccination was introduced, and the district school broke up for vacation. Since this constitutes a major change in the environment, we will only look at the first four generations.

The estimate $\hat{m}_4$ is 2.10. In order to use Theorem 2.3 we need to estimate σ^2 as well. Anticipating the next chapter, we will use the estimator

$$\hat{\sigma}^2 = \frac{1}{n-1} \sum_{i=1}^{n} Z_{i-1}(\bar{m}_i - \hat{m}_n)^2 ,$$

which in this case come out as 10.20. An approximate 95% confidence interval (albeit n is extremely small) is given by

$$\hat{m} \pm \frac{1.96\hat{\sigma}}{Y_{n-1}^{1/2}} ,$$

which in this case comes out to be (1, 3.47), since we are assuming $m > 1$. Using the rough guide derived in section 1 of this chapter, we would expect to need a randomly vaccinate $1 - 1/\hat{m}_4$, or 52% of the population. An upper confidence bound on this proportion, computed from the upper confidence bound on m, is 71%. The lower bound does not carry any information in this case. Although the confidence bound is likely to be conservative, since vaccination of well-chosen subsets of the population will be more effective than random vaccination, the data strongly suggest that mass vaccination was called for. □

When $m < 1$, conditional asymptotic normality fails, regardless of the norming used. We give the result of $\bar{m}_n$.

Proposition 2.8: Suppose that $m<1$ and $\sigma^2<\infty$. Then

$$\mathbf{P}\{\bar{m}_n = x | Z_{n-1} > 0\} \rightarrow \sum_{k\geq 1} p_{xk}^{*k} q_k ,$$

where (q_k) is the distibution with pgf $Q(s)$, given in Proposition 2.6.

Proof: Use that

$$\mathbf{P}(Z_n = j|Z_{n-1} > 0) = \sum_{k\geq 1} \mathbf{P}(Z_n = j|Z_{n-1} = k)\mathbf{P}(Z_{n-1} = k|Z_{n-1} > 0)$$

together with Proposition 2.6. □

Example: The Geometric Distribution

Using the results in the previous section for the geometric subcritical case, we see that $q_k = (2p-1)(1-p)^{k-1}p^{-k}$, while p_{kx}^{*k} is the probability function for a negative binomial distribution. It follows that the limiting conditional distribution of $\bar{m}_n$ can be written

$$\lim_{n\to\infty} \mathbf{P}\{\bar{m}_n = x|Z_{n-1} > 0\} = (2p-1) \sum_{h\geq 1, kx\in \mathbb{N}} \binom{kx-1}{k-1} (1-p)^{kx-1} .$$

For example, $\mathbf{P}\{\bar{m}_n = 1|Z_{n-1} > 0\} \rightarrow 2 - 1/p$ for $p > \frac{1}{2}$. □

A different kind of asymptotics is achieved by combining the two previous methods, i.e., to let initial sample size and time get large simultaneously. Denote an estimator at time n based on z initial individuals by $\bar{m}_n(z) = (Y_n(z) - z)/Y_{n-1}(z)$. The behavior turns out to be different on different parts of the parameter space. We need a few technical results.

Proposition 2.9: Suppose $\sigma^2 < \infty$. Then

(i) if $m<1$, $Y_n \xrightarrow[\text{a.s.}]{L_2} Y$ as $n\to\infty$, where $\mathbf{Var}\, Y = \sigma^2(1-m)^{-3}$;
(ii) if $m=1$ and $z\to\infty$, then if $n/z\to 0$, $Z_n/z \xrightarrow{\mathbf{P}} 1$; if $n/z\to\infty$, $Z_n/z \xrightarrow{\mathbf{P}} 0$; if $n/z \to K$, $0<K<\infty$, $Z_n/z \xrightarrow{d} U$, where U has density

$$f(u) = \frac{e^{-a(u+1)}}{a\sqrt{x}}\, I_1(2a\sqrt{x}) ,$$

$a = 2/(\sigma^2 K)$ and

$$I_1(x) = \sum_{k=1}^{\infty} \frac{1}{(k-1)!k!} (x/2)^{2k-1} .$$

This is the density of the sum of a Poisson(a) number of independent exponential variates, each with mean $1/a$.

Proof: To verify (i), notice that

$$\mathbf{E}Z_kZ_l = m^{l-k}\mathbf{E}Z_k^2,$$

so that

$$\begin{aligned}\mathbf{E}Y_n^2 &= \sum \mathbf{E}Z_k^2 + 2\sum_{k<l} \mathbf{E}Z_kZ_l \\ &= \sigma^2\left\{\frac{(1-m^n)(1+m^{n+1})}{(1-m)^3} - \frac{2nm^n}{(1-m)^2}\right\}.\end{aligned}$$

It follows that $\mathbf{E}(Y_{k+n} - Y_n)^2 \to 0$ as $k \to \infty$. Hence the L_2-convergence and the variance expression. The a.s. convergence follows since $Z_n \to 0$ a.s. and the Z_n are integer valued, so that for almost all ω the number of nonzero terms in Y_∞ is finite.

To prove (ii), we use this result.

Lemma 2.4: If $m = 1$

$$1 - P_n(s) = \frac{1-s}{1+\sigma^2 n(1-s)/2}(1+o(1))$$

as $n \to \infty$. The error term is uniform in $|s| \le 1$.

Proof of Lemma: Let $R_n(s) = 1 - P_n(s)$. Since $P_{n+1}(s) = P(P_n(s))$ we get by Taylor expansion of $P(s)$ about $s = 1$ that

$$R_{n+1}(s) = R_n(s) - \frac{\sigma^2}{2}R_n^2(s) + O(R_n^3(s)). \tag{8}$$

Notice that

$$|R_n(s)| \le 1 - P_n(0) + (P_n(s) - P_n(0)) \le 2(1-q_n), \tag{9}$$

where $q_n = P_n(0)$. Combining (8) and (9) we see that

$$\left|\frac{R_n(s)}{R_{n+1}(s)} - 1\right| \le c(1-q_n)$$

uniformly in s. Dividing (8) by $R_n(s)R_{n+1}(s)$ and denoting $w_n(s) = 1/R_n(s)$ we get

$$w_{n+1}(s) = w_n(s) + \frac{\sigma^2}{2} + O(1-q_n).$$

By summing on n, noticing that the sum telescopes and that $P_0(s) = s$, we get

$$w_n(s) = \frac{1}{1-s} + n\left(\frac{\sigma^2}{2} + o(1)\right),$$

since $q_n \to 1$ and $n \to \infty$, and the error term is uniform in s. □

Proof of Proposition (continued): The Laplace transform of $Z_n(z)/z$ is $P_n(e^{-\lambda/z})^z$. From the Lemma we get

$$-\log P_n(e^{-\lambda/z})^z \sim \frac{\lambda}{1 + \lambda\sigma^2 n/2z}$$

so that

$$P_n(e^{-\lambda/z})^z = \exp\left(-\frac{\lambda}{1 + \lambda\sigma^2 n/2z}\right)(1 + o(1)).$$ □

Corollary: $1 - q_n \sim 2/(n\sigma^2)$ when $m = 1$.

Proof: Set $s = 0$ in Lemma 2.4 □

Remark: When $m > 1$, $q - q_n = dP'(q)^n + o(P'(q)^n)$ for some positive d. When $m < 1$, $1 - q_n = O(m^n)$. See Exercise 9. □

Application: Survival of Gene Mutations

Consider a plant that has a large number of offspring, such as a cross-pollinated cereal plant. Each sexually mature individual is the mother of a large number, n, of mature grains, and the father, on average, of an equal number. If we consider a particular mutated gene, it has probability $\frac{1}{2}$ to enter each of these grains. But on the average only two grains will be represented in next year's crop (assuming no selective advantage). The overall probability for the gene to appear in the next generation is then about $1/2n$, and the number of surviving genes is binomially distributed if we assume independence. Approximately, then, the number of surviving genes will have a Poisson distribution with parameter 1. Variability in n may lead to overdispersion relative to the Poisson, which has sometimes been observed (e.g., Kojima and Kelleher, 1962). By the Corollary, the survival probability is about $2/n$ for large n. If on the other hand the gene in question is increasing in frequency by a factor of c each generation, we can argue similarly that the distribution of the number in the next generation should be approximately Poisson with mean c. Using Exercise 9, we see that the survival probability now approaches $1 - q$ exponentially fast, as opposed to the linear rate of the Corollary. The following table gives $1 - q_n$ for some values of n.

n	$c=1$	$c=1.01$
1	0.632	0.636
127	0.015	0.027
∞	0.000	0.020

For mutations with a 1% selective advantage and which each exist in one sexually mature plant, on average 36% will fail to transmit the gene to any descendant. Without any advantage, still only 37% will fail. After 127 generations, 1.5% survive on average without the advantage, 2.7% with it. In other words, although a selective advantage of 1% would be extremely powerful when present in large numbers, this advantage is almost negligible compared to chance survival if it is only present in a few individuals. On the other hand, since a mutation with a 1% selective advantage has a 2% probability of spreading throughout the species, such mutations must be extremely rare. If they occurred with even moderate frequency, they would soon be successful. Typical mutation rates are smaller than 10^{-6}. □

We are now ready to see what happens to $\hat{m}_n(z)$ when both n and z get large.

Theorem 2.4

(i) As $n, z \to \infty$, $\hat{m}_n(z) \xrightarrow{\mathbf{P}} m$ for all m.

(ii) If $m < 1$, as $n, z \to \infty$

$$(\sigma^2(1-m)/z)^{1/2}(\hat{m}_n(z) - m) \xrightarrow{d} \mathbf{N}(0, 1)\,.$$

(iii) If $m = 1$, as $n, z \to \infty$ with $n/z \to 0$

$$(\sigma^2/nz)^{1/2}(\hat{m}_n(z) - 1) \xrightarrow{d} \mathbf{N}(0, 1)\,,$$

whereas if $n/z^2 \to \infty$

$$\frac{\sigma^2}{2z^2(1-\hat{m}_n(z))} \xrightarrow{d} Y\,,$$

where Y has a stable law of index $\frac{1}{2}$.

(iv) If $m > 1$, as $n, z \to \infty$

$$\left(\frac{\sigma^2 Z(m-1)}{z(m^n - 1)}\right)^{1/2} (\hat{m}_n(z) - m) \xrightarrow{d} \mathbf{N}(0, 1)\,.$$

Proof: (i) Let $m<1$. By Proposition 2.9(i) and the law of large numbers

$$Y_n(z)/z \xrightarrow{\mathbf{P}} 1/(1-m)\,.$$

Now write

$$\hat{m}_n(z) = (Y_n(z)/z - 1)/(Y_{n-1}(z)/z)\,.$$

If $m=1$, write

$$\hat{m}_n(z) = 1 + (Z_n(z)/z - 1)/(Y_{n-1}(z)/z)\,,$$

and check (Exercise 12) that $Y_{n-1}(z)/z \to \infty$ in probability whenever n/z does not converge to zero. When $n/z \to 0$, use Proposition 2.9(ii). Finally, if $m>1$ we use the Corollary to Theorem 2.3 to get

$$U_n = \frac{m-1}{m^n-1}\,Y_{n-1} \underset{\text{a.s.}}{\overset{L_2}{\to}} W\,. \tag{10}$$

Now write

$$\hat{m}_n(z) = \frac{\dfrac{m-m^{-n}}{1-m^{-n}}\,\dfrac{1}{z}\displaystyle\sum_{k=1}^{2} U_{n+1}(k) - \dfrac{m-1}{m^n-1}}{\dfrac{1}{z}\displaystyle\sum_{k=1}^{2} U_n(k)}\,,$$

where the $U_n(i)$ are iid as U_n, and recall that $\mathbf{E}W=1$.

To prove (ii), look at $w_k^{(i)} = Z_k^{(i)} - mZ_{k-1}^{(i)}$, and let $M_n^{(i)} = \Sigma_k\, w_k^{(i)}$. Each $M_n^{(i)}$ is an L_2-martingale, converging a.s. to some $M^{(i)}$. Write

$$\hat{m}_n(z) - m = Y_{n-1}^{-1}(z)\sum_{i=1}^{z}\sum_{k=1}^{n} w_k^{(i)}$$

so that

$$\frac{\sqrt{z}}{\sigma\sqrt{1-m}}\,(\hat{m}_n(z)-m) = \frac{\dfrac{1}{\sqrt{z}}\displaystyle\sum_{i=1}^{z}(M_n^{(i)} - M^{(i)})}{\dfrac{1}{z}\,Y_{n-1}(z)\sigma\sqrt{1-m}} + \frac{\dfrac{\sqrt{1-m}}{\sigma\sqrt{n}}\displaystyle\sum_{i=1}^{z} M^{(i)}}{\dfrac{1}{n}\,Y_{n-1}(z)(1-m)}\,.$$

The first term goes in probability to zero by Chebyshev's inequality, and the second is asymptotically normal by the central limit theorem.

In order to verify the first statement of (iii), we first note that

$$Y_{n-1}(z)/nz \xrightarrow{\mathbf{P}} 1\,,$$

since $\mathbf{E}Y_{n-1}(z) = nz$ and $\mathbf{Var}\ Y_{n-1}(z) = z\sigma^2(n-1)n(2n-1)/6$. Using Lemma 2.4 we can prove that

$$\left(\frac{z}{\sigma^2 n}\right)^{1/2} \left(\frac{Z_n(z)}{z} - 1\right) \xrightarrow{d} \mathbf{N}(0, 1)\,,$$

since the characteristic function of this random variable is

$$\exp\left(-\frac{i\lambda}{\sigma}\left(\frac{z}{n}\right)^{1/2}\right) P_n(e^{i\lambda/\sigma\sqrt{zn}}) = e^{-\lambda^2/2}(1 + o(1))\,.$$

Now note that

$$\frac{\sqrt{nz}}{\sigma}\,(\hat{m}_n(z) - 1) = \frac{\left(\frac{z}{\sigma^2 n}\right)^{1/2}(Z_n(z)/z - 1)}{Y_{n-1}(z)/nz}\,.$$

The second statement of (iii) follows from Proposition 2.9(ii) and the fact that

$$\frac{\sigma^2}{2z^2}\, Y_n(z) \xrightarrow{d} Y\,,$$

To prove this fact, we write the logarithm of the Laplace transform of the left-hand side:

$$\begin{aligned}\log \mathbf{E}e^{-\lambda\sigma^2 Y_n(z)/2z^2} &= -z(1 - R(e^{-\lambda\sigma^2/2z^2})) + z(R_n(e^{-\lambda\sigma^2/2z^2}) - R(e^{-\lambda\sigma^2/2z^2}))\\ &= \sqrt{\lambda} + o(n/z^2) + (R_n(e^{-\lambda(\sigma/z)^2}) - R(e^{-\lambda(\sigma/z)^2}))\,, \qquad (11)\end{aligned}$$

where R_n is the pgf of Y_n^1, R is the pgf of Y_∞ and we used the hint to Exercise 12. Since

$$R_n(s) = sP(R_{n-1}(s))\,,$$

a Taylor expansion of $P(s)$ shows that $R_n(s) = s^{n+1} + o(s^{n+1})$. Since $R(s) \leq s^n R(s)$, the last term on the right-hand side of (11) goes to zero in probability.

Finally, in order to prove (iv), introduce the variable

$$V_{n,z} = \frac{M_n}{\sqrt{z\ \mathbf{Var}\ M_n}}$$

where $M_n = \Sigma_1^n (Z_k - mZ_{k-1})$, and let $V_{n,z}^{(i)}$, $i = 1, \ldots, z$ be iid copies of $V_{n,z}$. Using the Lindeberg criterion we must verify that

$$z \int_{|x| \geq \epsilon} x^2\, d\mathbf{P}(V_{n,z} \leq x) \rightarrow 0$$

as $n, z \to \infty$. Let $V_n = \sqrt{z} V_{n,z}$. It suffices to show that

$$\mathbf{E}(V_n^2 1(|Z_n| \geq \epsilon\sqrt{z}) \to 0 .$$

By the Cauchy–Schwarz and Chebyshev inequalities, it suffices to prove that $\mathbf{E}V_n^4 = O(1)$. A messy computation shows that $\mathbf{E}M_n^4 = O(m^{2n})$, whence the result follows. Finally write

$$\left(\frac{z(m^n - 1)}{m - 1}\right)^{1/2} (\hat{m}_n(z) - m) = \left(\frac{1}{z} \sum_{i=1}^{z} U_n^{(i)}\right)^{-1} \sum_{i=1}^{z} V_{n,z}^{(i)} ,$$

where the $U_n^{(i)}$ are iid as the U_n of (10). □

Example: Queueing Theory

Another approach to limit theory for BGW processes obtains in a queueing example. Suppose that we study a M/G/1-queue, starting with one individual arriving and being served at time 0. Consider the number of customers arriving during the service time of a customer as his offspring. This defines a BGW process with offspring probabilities

$$p_j = \int_0^\infty (\lambda t)^j e^{-\lambda t}/j!\, dF(t) ,$$

where $F(x)$ is the service time distribution. The offspring mean is $\lambda\mu$ where μ is the mean of the service time distribution. If the service station becomes empty, we get an independent realization of the process starting with the next customer. If, in addition, the service station has limited queue length, and closes down to clear the service area whenever the queue length exceeds this limit, the process becomes a bounded process. Whenever the process becomes extinct or closes down we get another independent realization of the process. Since the process is bounded, so that the asymptotic normality of large generations is avoided, we may be able to get consistent estimates of the offspring distribution. Eschenbach and Winkler (1975) discussed asymptotics in this setup, and obtained consistency of offspring probability estimators in the case of offspring distributions with finite support. This process is, in fact, an ergodic Markov chain, so the results of Billingsley (1962) apply. □

7. EFFICIENCY AND EXPONENTIAL FAMILIES

The intuitive meaning of an efficient estimator is one that draws maximal information out of the observations. In probability terms, the estimator must have a distribution which is as concentrated as possible about the true parameter value. Typically, even in the iid case, this can only be achieved in an asymptotic sense, unless one restricts attention to a subclass of estimators, such as the unbiased ones. Based on the principle that the likelihood contains all the relevant information about the underlying distribution (see Cox and Hinkley, 1974, section 2.3, for a discussion of this principle), we will restrict attention to Rao's approach to efficiency, which is to look at estimators with maximum correlation with the normalized likelihood.

We will begin by studying the likelihood martingale. For now, we will be discussing general concepts and not the branching process case per se. Let $X_1, \ldots, X_n$ be observations from a stochastic process with distribution depending on an unknown parameter θ. Assume that θ takes values in some open subset Θ of the real line. Let $p_n(x_1, \ldots, x_n; \theta)$ be the density of $(X_1, \ldots, X_n)$ with respect to some σ-finite measure ν. We will assume the Cramér type condition that

$$\int p_n(x_1, \ldots, x_n; \theta)\nu\,(dx_1, \ldots, dx_n)$$

can be differentiated twice with respect to θ under the integral sign. Define the likelihood $L_n(\theta) = p_n(x_1, \ldots, x_n; \theta)$ and let $\mathcal{F}_k = \sigma(X_1, \ldots, X_k)$ be the natural history. Also, let $\mathcal{F}_0 = \{\emptyset, \Omega\}$ and $L_0 = 1$. Define

$$u_i(\theta) = \frac{d}{d\theta} \log p_1(x_1, \ldots, x_i | \mathcal{F}_{i-1}; \theta)\,.$$

Then

$$M_n(\theta) = \frac{d \log L_n(\theta)}{d\theta} = \sum_{i=1}^{n} u_i(\theta)\,,$$

where $\mathbf{E}_\theta(u_i(\theta)|\mathcal{F}_{i-1}) = 0$, so $(M_n, \mathcal{F}_n, n \geq 0)$ is a square integrable martingale. Define the *conditional information*

$$I_n(\theta) = \sum_{i=1}^{n} \mathbf{E}_\theta(u_i^2(\theta)|\mathcal{F}_{i-1})\,.$$

Note that by interchanging integration and differentiation

$$\mathbf{E}_\theta(u_i^2(\theta)|\mathcal{F}_{i-1}) = \mathbf{E}_\theta\left(-\frac{du_i(\theta)}{d\theta}\middle|\mathcal{F}_{i-1}\right).$$

Write $J_n(\theta) = \Sigma_{i=1}^{n} \nu_i(\theta)$, where $\nu_i(\theta) = -d(u_i(\theta))/d\theta$. $J_n(\theta)$ is called the *observed information*. Clearly $(I_n(\theta) - J_n(\theta), \mathcal{F}_n, n \geq 1)$ is a martingale. A related quantity is the *Fisher information* $\xi_n(\theta) = \mathbf{E}_\theta I_n(\theta) = \mathbf{E}_\theta J_n(\theta)$. The three quantities $I_n(\theta)$, $J_n(\theta)$, and $\xi_n(\theta)$ all measure aspects of the information about θ contained in a sample.

Example: The iid Exponential Family

These quantities are all well known in the iid case. Assume that we are studying a one-parameter exponential family, so that

$$L_n(\theta) = \exp\left[c(\theta) \sum_1^n T(x_i) + nd(\theta) \right].$$

Then $u_i(\theta) = c'(\theta)T(x_i) + d'(\theta)$ so

$$I_n(\theta) = -c''(\theta) \sum \mathbf{E}T(X_i) - nd''(\theta) = n\left(\frac{c''(\theta)d'(\theta)}{c'(\theta)} - d''(\theta) \right)$$

is the Fisher information and

$$J_n(\theta) = -c''(\theta) \sum T(x_i) - nd''(\theta)$$

is the observed information (see the discussion in Efron and Hinkley, 1978, as to why the latter is a better estimate of the asymptotic variance of mles; another argument can be based on Proposition 4.3). If the exponential family is in its natural form, i.e., parametrized using $\eta = c(\theta)$, the observed information and the Fisher information about η are the same. □

Example: Branching Process with Power Series Offspring Distribution

The power series offspring distribution is given in Exercise 6. A simple computation shows that

$$M_n(m) = \frac{1}{\sigma^2(m)} (Y_n - Z_0 - mY_{n-1}).$$

Since

$$u_k = (Z_k - mZ_{k-1})/\sigma^2(m),$$

we have that

$$I_n(m) = \sum_{i=1}^n \mathbf{E}(u_i^2 | \mathcal{F}_{i-1}) = Y_{n-1}/\sigma^2(m).$$

Furthermore,

$$v_n(m) = \frac{d}{dm} \sigma^2(m) \frac{Z_n - mZ_{n-1}}{\sigma^4(m)} + \frac{Z_{n-1}}{\sigma^2(m)}$$

so

$$J_n(m) = I_n(m) + \frac{d}{dm} \log(\sigma^2(m)) M_n(m).$$

Finally

$$\xi_n(m) = \sigma^2(m)\,\frac{m^n - 1}{m - 1}\,. \qquad \square$$

Under mild regularity conditions we can prove the existence of a consistent root of the likelihood equation.

Proposition 2.10: Suppose that $I_n(\theta) \to \infty$ a.s. Then

$$I_n^{-1}(\theta) M_n(\theta) \xrightarrow{\text{a.s.}} 0\,,$$

and if also

$$\limsup_{n\to\infty} \frac{|I_n(\theta) J_n(\theta_n^*)|}{I_n(\theta)} < 1 \text{ a.s.}$$

whenever $|\theta_n^* - \theta| < \delta$ for some $\delta > 0$, there is a consistent root $\hat{\theta}$ of the likelihood equation

$$M_n(\theta) = 0\,.$$

Proof: For the first part we shall use the following martingale strong law (Hall and Heyde, 1980, Theorem 2.18).

Lemma 2.5: Let $(\zeta_n = \Sigma_{i=1}^n \xi_i, \mathscr{F}_n, n \ge 1)$ be a martingale, and let $(\mu_n, n \ge 1)$ a nondecreasing sequence of positive random variables, such that $\mu_n \in \mathscr{F}_{n-1}$. Then $\mu_n^{-1}\zeta_n \to 0$ a.s. on the set

$$\left\{\lim \mu_n = \infty,\ \sum_{i=1}^{\infty} \mu_i^{-2} \mathbf{E}(\xi_i^2 | \mathscr{F}_{i-1}) < \infty\right\}.$$

Using $\mu_n = I_n(\theta)$, we must show that

$$\sum_{n=1}^{\infty} I_n(\theta)^{-2} \mathbf{E}(u_n^2 | \mathscr{F}_{n-1}) < \infty \text{ a.s.}$$

But this is immediate from the following result.

Lemma 2.6: Let $a_1, a_2, \ldots$ be a sequence of positive real numbers, and set $A_k = \Sigma_1^k a_i$. Then

$$\sum_2^{\infty} \frac{a_i}{A_i^2} \le a_1^{-1}\,.$$

Proof of Lemma 2.6: Without any loss of generality we can let $a_1 = 1$. Write

$$\sum_2^{\infty} \frac{a_i}{A_i^2} = \sum_2^{\infty} \frac{A_i - A_{i-1}}{A_i^2} \le \int_1^{\infty} \frac{dx}{x^2} = 1\,. \qquad \square$$

Proof of Proposition (continued): For the second part, expand around the true parameter value θ_0:

$$M_n(\theta) = M_n(\theta_0) + (\theta - \theta_0) \sum \nu_i(\theta_n^*)$$
$$= M_n(\theta_0) - (\theta - \theta_0) I_n(\theta_0) + (\theta - \theta_0)(I_n(\theta_0) - J_n(\theta_n^*)) , \qquad (12)$$

where $\theta_n^* = \theta_0 + \gamma(\theta - \theta_0)$ for some $\gamma = \gamma(n, \theta_0)$ with $|\gamma| < 1$. Solving the likelihood equation $M_n(\theta) = 0$ yields

$$(\hat{\theta} - \theta_0)\left(1 - \frac{I_n(\theta_0) - J_n(\theta_n^*)}{I_n(\theta_0)}\right) = \frac{M_n(\theta_0)}{I_n(\theta_0)} ,$$

whence the claim follows, using the first part of the proposition. □

Example: The iid Exponential Family

In the iid case of an exponential family in the natural parametrization, the condition of Proposition 2.10 translates into

$$|d''(\eta^*) - d''(\eta)| < |d''(\eta)|$$

whenever $|\eta^* - \eta| < \delta$; in other words, a growth condition on d''. A Taylor expansion shows that if d has a bounded third derivative, this condition is satisfied whenever d'' is bounded below in a neighborhood of η, which can be obtained whenever $d''(\eta) \neq 0$ by the convexity of d. Lemma 2.5 with $\mu_n = n$ yields that $M_n(\eta)/n \to 0$ a.s. whenever $\Sigma\, i(\eta)/n^2 < \infty$, i.e., whenever $i(\eta) < \infty$, where $i(\eta)$ is the Fisher information in a single observation. □

Example: Branching Process with Power Series Offspring Distribution

First write

$$\frac{I_n(m) - J_n(m^*)}{I_n(m)} = \frac{I_n(m) - J_n(m)}{I_n(m)} + \frac{J_n(m) - J_n(m^*)}{I_n(m)} .$$

Using the computations in the example preceding Proposition 2.10 we can write the first term as

$$-\frac{d}{dm} \log \sigma^{-2}(m) \frac{M_n(m)}{I_n(m)} ,$$

which converges a.s. to 0 on the explosion set by Proposition 2.10, while the second term is

$$\frac{I_n(m) - I_n(m^*)}{I_n(m)} - \frac{d}{dm} \log \sigma^{-2}(m^*) \frac{M_n(m^*)}{I_n(m^*)} \frac{I_n(m^*)}{I_n(m)}$$
$$+ \frac{d}{dm} \log \sigma^{-2}(m) \frac{M_n(m)}{I_n(m)} = \left(1 - \frac{\sigma^2(m)}{\sigma^2(m^*)}\right) + o_{\text{a.s.}}(1)$$

on the explosion set, again using the Proposition 2.10. This can be made arbitrarily close to zero by the continuity of $\sigma^2(m)$. □

It is frequently the case that

$$I_n(\theta)^{-1/2} M_n(\theta) \xrightarrow{d} \mathbf{N}(0, 1) , \qquad \text{continuously in } \theta \in \Theta , \tag{13}$$

in the sense that whenever θ_1 and θ_2 are close, the distribution functions of the corresponding random variables are close for n large enough. Hence the convergence is uniform on compact sets of θ-values. Such uniformity is necessary to produce asymptotically correct confidence intervals: if the convergence is not continuous in θ it is not possible to invert the asymptotic pivot $I_n^{-1/2} M_n$.

Example: Branching Process with Power Series Offspring Distribution

The continuous convergence is a consequence of the remark following Theorem 2.3. □

Proposition 2.11: Assume that (12) holds, that

$$\frac{J_n(\theta)}{I_n(\theta)} \xrightarrow{\mathbf{P}_\theta} 1 \text{ uniformly for } \theta \in S_n(\theta_0) = \left\{\theta \colon |\theta - \theta_0| < \frac{\delta}{\xi_n(\theta_0)^{1/2}}\right\},$$

where $\delta > 0$ is a fixed number, and that

$$I_n(\theta_0)^{-1}(J_n(\theta) - J_n(\theta_0)) \to 0 \text{ a.s.}$$

for any $\theta \in S_n(\theta_0)$. Then

$$I_n(\theta_0)^{1/2}(\hat{\theta}_n - \theta_0) \xrightarrow{d} \mathbf{N}(0, 1)$$

continuously in θ.

Proof: Use the expansion (11), and verify that the conditions given imply that

$$\frac{J_n(\theta_n^*)}{I_n(\theta_0)} \to 1 \text{ a.s.}$$ □

Remark: The conditions (except (13)) in Proposition 2.11 ensure that the second condition of Proposition 2.10 holds. □

Example: Branching Process with Power Series Offspring Distribution

The computations in the Examples above show that, in fact, the conditions of Proposition 2.11 are satisfied. □

If $\hat{\theta}$ is to be efficient among all estimators that achieve asymptotic normality at the same rate, any other estimator must have an asymptotic variance of at least 1. In this vein, say that an estimator T_n is *Rao-efficient* if

$$I_n(\theta)^{1/2}\{T_n - \theta - \gamma(\theta)I_n(\theta)^{-1}M_n(\theta)\} \xrightarrow{\mathbf{P}_\theta} 0 . \tag{14}$$

This is quite similar to Rao's definition of asymptotic first order efficiency for iid models. The idea is that a first order efficient estimator should asymptotically be a linear function of the derivative of the log likelihood, since the latter contains all the information available from the data about θ. A first order efficient estimator is consistent, and the definition suggests considering the best Rao-efficient estimator, in the sense of achieving the smallest $\gamma(\theta)$. If the convergence in (14) is strengthened to L_2-convergence, we see that an equivalent formulation is

$$\mathbf{corr}(I_n(\theta)^{1/2}(T_n - \theta), I_n(\theta)^{-1/2}M_n(\theta)) \to 1 .$$

Under condition (13),

$$I_n(\theta)^{1/2}(T_n - \theta) \xrightarrow{d} \mathbf{N}(0, \gamma(\theta)^2) .$$

Provided the conditions of Proposition 2.11 are satisfied, we would say that T_n has asymptotic relative efficiency of $\gamma(\theta)^{-2}$ with respect to the mle $\hat{\theta}_n$. This statement is interesting if we can verify that $\gamma(\theta) \leq 1$, at least under reasonable regularity conditions, since then the mle would be optimal among Rao-efficient estimators in the sense of achieving the lowest possible asymptotic variance. The notion of relative efficiency is a second order notion, so that two first order efficient estimators may have relative efficiency different from one. Continuity of the convergence in distribution is not only important for constructing confidence intervals, but it is also one way to eliminate superefficient estimators, for which $\gamma(\theta) < 1$. We will be looking at local alternatives, i.e, alternatives close ot the true parameter value θ_0, and define

$$\theta_n = \theta_0 + \frac{h}{(\xi_n(\theta_0))^{1/2}} .$$

We will see that in this context $\xi_n(\theta_0)$ plays the role of the sample size in the iid setup. In many cases we have that

$$\frac{I_n(\theta)}{\mathbf{E}_\theta I_n(\theta)} \xrightarrow{\mathbf{P}_\theta} \eta(\theta) \tag{15}$$

for some random variable η. In the ergodic case η is identically 1.

Theorem 2.5: Suppose that

$$I_n(\theta)^{1/2}(T_n - \theta) \xrightarrow{d} \mathbf{N}(0, \gamma(\theta)^2) \qquad \text{(mixing)},$$

uniformly on compact subsets of Θ, where $\gamma(\theta)$ is a bounded function. Assume that

$$\xi_n(\theta_0)^{-2}\mathbf{E}_\theta(I_n(\theta)^2 - J_n(\theta))^2 \tag{16}$$

is bounded uniformly over $\{\theta: |\theta - \theta_0| \le h/\xi_n(\theta_0)^{1/2}\}$ for any Borel set $B_n \in \mathcal{F}_n$. Then, under the assumptions of Proposition 2.11 and if (15) holds, we have that $\gamma^2(\theta) \ge 1$.

Proof: We will look at the locally most powerful test of $\theta = \theta_0$ against $\theta > \theta_0$, which is the test with the largest slope of the power function at θ_0. Then we will compare that to a competitor based on t_n, which must have no larger slope. The LMP test rejects when $\{\Delta_n(\theta_0) > r_n(\alpha)\}$ (Rao, 1973, section 7a.4; see also section 10), where $\Delta_n(\theta) = M_n(\theta)/(\xi_n(\theta_0))^{1/2}$ and $r_n(\alpha)$ is the α-quantile of the distribution of Δ_n. Since $\Delta_n(\theta_0) \xrightarrow{d} \eta(\theta_0)^{1/2}Z$, where $Z \sim \mathbf{N}(0, 1)$, we see that $r_n(\alpha) \to r(\alpha)$, the upper α-point of the distribution of $\eta(\theta_0)^{1/2}Z$. Define the power function

$$\pi_n(\theta_n) = \mathbf{P}_{\theta_n}\{\Delta_n(\theta_0) > r_n(\alpha)\} = \mathbf{P}_{\theta_n}\{\Delta_n(\theta_n) + \Delta_n(\theta_0) - \Delta_n(\theta_n) > r_n(\alpha)\}.$$

Verify that $\Delta_n(\theta_n) - \Delta_n(\theta_0) \xrightarrow{\mathbf{P}_\theta} \eta(\theta_0)h$, so

$$\lim_{n\to\infty} \pi_n(\theta_n) = \mathbf{P}(\eta(\theta_0)Z \ge r(\alpha) - \eta(\theta_0)h).$$

By Taylor expansion we see that

$$\frac{1}{h}(\pi_n(\theta_n) - \pi_n(\theta_0)) = (\xi_n(\theta_0))^{-1/2}\pi_n'(\theta_0) + o(h).$$

The last two equations yield that

$$\begin{aligned}\limsup (\xi_n(\theta))^{-1/2}\pi_n'(\theta_0) &= \lim_{h\downarrow 0}\limsup_{n\to\infty} \frac{1}{h}(\pi_n(\theta_n) - \pi_n(\theta_0)) \\ &= \lim_{h\downarrow 0}\frac{1}{h}\mathbf{E}_{\theta_0}(\Phi(\eta^{-1/2}r(\alpha)) \\ &\quad - \Phi(\eta^{-1/2}r(\alpha) - \eta^{1/2}h)) \\ &= (2\pi)^{-1/2}\mathbf{E}_{\theta_0}(\eta^{1/2}e^{-r^2(\alpha)/2\eta}), \end{aligned} \tag{17}$$

where $\eta = \eta(\theta_0)$.

Now contrast the LMP test to a test based on T_n, rejecting when

$$(\xi_n(\theta_0))^{-1/2}I_n(\theta_0)(T_n - \theta_0)\gamma(\theta_0)^{-1} > \rho_n(\alpha).$$

The random quantity on the left-hand side of the inequality converges uniformly in distribution to $\eta^{1/2}Z$, so $\rho_n(\alpha) \to r(\alpha)$. The power of this test at θ_n is

$$\begin{aligned}\pi_n^*(\theta_n) &= \mathbf{P}\left(T_n - \theta_n \ge \theta_0 - \theta_n + \frac{\xi_n(\theta_0)^{1/2}}{I_n(\theta_0)}\rho_n(\alpha)\gamma(\theta_0)\right) \\ &= \mathbf{P}\left(\frac{I_n(\theta_n)}{(\xi_n(\theta_n))^{1/2}\gamma(\theta_0)}(T_n - \theta_n)\right. \\ &\quad \left.\ge \frac{I_n(\theta_n)}{(\xi_n(\theta_n))^{1/2}\gamma(\theta_0)}(\theta_0 - \theta_n) + \left(\frac{\xi_n(\theta_0)}{\xi_n(\theta_n)}\right)^{1/2}\frac{I_n(\theta_n)}{I_n(\theta_0)}\rho_n(\alpha)\right) \\ &\to \mathbf{P}\left(\eta^{1/2}Z \ge -h\frac{\eta}{\gamma(\theta_0)} + r(\alpha)\right).\end{aligned}$$

Thus, arguing as in the LMP case,

$$\limsup_{n\to\infty}(\xi_n(\theta_0))^{-1/2}\pi_n^{*\prime}(\theta_0) = (2\pi)^{-1/2}\gamma(\theta_0)^{-1}\mathbf{E}_{\theta_0}(\eta^{1/2}e^{-r^2(\alpha)/2\eta}). \tag{18}$$

Since $\pi_n'(\theta_0) \ge \pi_n^{*\prime}(\theta_0)$ we see by comparing (17) to (18) that $\gamma(\theta_0) \ge 1$. □

Remark: Theorem 2.5 is valid under the weaker condition $\xi_n^{-1}\int_{B_n}(d^2L_n(\theta)/d\theta^2)\,d\mathbf{P}$ uniformly bounded in local neighborhoods of θ for all Borel sets B_n, replacing (16). However, in particular cases (16) appears easier to check. □

Under the same regularity conditions we see that any two first order efficient estimators T_n and U_n with asymptotic relative efficiency 1 are equivalent, in the sense that

$$\begin{aligned}I_n^{1/2}(\theta)(T_n - U_n) &= I_n^{1/2}(\theta)(T_n - \theta - \gamma(\theta)I_n^{-1}M_n - U_n + \theta + \gamma(\theta)I_n^{-1}M_n) \\ &\xrightarrow{\mathbf{P}} 0,\end{aligned}$$

which implies that one would use the same asymptotic confidence statements for T_n and U_n.

Notice that in the efficiency definition (14) we are using the random norming $I_n(\theta)$. In other words, we are letting the particular realization decide on the "right" speed of convergence. This is appropriate for estimation theory (cf. the discussion in section 9), but is inappropriate if we are aiming to plan an experiment involving a stochastic process. A fixed norming, which then will have to be $\xi_n(\theta_0)$, would be needed in that situation. Rao-efficiency is strongly geared towards asymptotic normality of the estimators, because of the asymptotic normality (13) of the score function, and if we do not use random normings, we may not have asymptotically normally distributed estimators. Weiss and Wolfowitz (1974, chapter 3) discuss a more general sense of efficiency which handles nonnormal limiting distributions. In this situation W-efficiency leads to comparisons of the kind

$$\lim_{n\to\infty} \mathbf{P}_{\theta_0}(-c < (\xi_n(\theta_0))^{1/2}(\hat{\theta}_n - \theta_0) < b)$$
$$\geq \limsup_{n\to\infty} \mathbf{P}_{\theta_0}(-c + a(\theta_0) < (\xi_n(\theta_0))^{1/2}(T_n - \theta_0) < A(\theta_0) + b)\,,$$

where $a(\theta) < A(\theta)$ and b, c are arbitrary positive numbers. This kind of inequality can be shown to hold under mild regularity conditions, provided $\eta(\theta) = 1$. However, in nonergodic cases we have to confine our attention to symmetric intervals of concentration. This is reasonable for estimators whose limiting distribution is a scale mixture of normals, and hence symmetric. Under continuity conditions on $I_n(\theta)$ and $J_n(\theta)$, similar to the ones used above, Hall and Heyde (1980) prove the following theorem.

Theorem 2.6: Let T_n be an estimator of θ satisfying

$$\lim_{n\to\infty}(\mathbf{P}_\theta((\mathbf{E}_\theta I_n(\theta))^{1/2}(T_n - \theta) > -\delta) - \mathbf{P}_{\theta_n}((\mathbf{E}_\theta I_n(\theta))^{1/2}(T_n - \theta_n) > -\delta)) = 0$$

for all $\theta \in \Theta$. Under suitable regularity conditions there exists a consistent mle $\hat{\theta}_n$, satisfying

$$\lim \mathbf{P}_{\theta_0}(|(\xi_n(\theta_0))^{1/2}(\hat{\theta}_n - \theta_0)| < \delta) \geq \limsup \mathbf{P}_{\theta_0}(|(\xi_n(\theta_0))^{1/2}(T_n - \theta_0)| < \delta)\,.$$

The proof relies on contiguity arguments and is omitted.

In order to use the criterion (14) in practice, we must have a tractable expression for M_n. The simplest case arises when

$$M_n(\theta) = I_n(\theta)(\hat{\theta}_n - \theta)\,. \tag{19}$$

In the iid case, this condition specifies an exponential family. In the general case it imposes a certain structure on the information.

Proposition 2.12: Equation (19) holds for all $n \geq 1$ if and only if

$$I_n(\theta) = \phi(\theta) H_n(X_1, \ldots, X_{n-1}) . \tag{20}$$

Proof: Assume first that (19) holds. Differentiating with respect to θ we get

$$M_n'(\theta) + I_n(\theta) = I_n'(\theta)(\hat{\theta}_n - \theta) ,$$

that is,

$$\frac{I_n'(\theta)}{I_n(\theta)} = \frac{M_n'(\theta) + I_n(\theta)}{M_n(\theta)} . \tag{21}$$

Now compute

$$\begin{aligned} \mathbf{E}(M_n'(\theta) | \mathcal{F}_{n-1}) &= I_n'(\theta) \mathbf{E}(\hat{\theta}_n - \theta | \mathcal{F}_{n-1}) - I_n(\theta) \\ &= \frac{I_n'(\theta)}{I_n(\theta)} M_{n-1}(\theta) - I_n(\theta) . \end{aligned}$$

Use that $\mathbf{E}(u_i(\theta)^2 | \mathcal{F}_{i-1}) = -\mathbf{E}(u_i'(\theta) | \mathcal{F}_{i-1})$ to verify that

$$\mathbf{E}(M_n'(\theta) | \mathcal{F}_{n-1}) = M_{n-1}'(\theta) - (I_n(\theta) - I_{n-1}(\theta)) ,$$

whence

$$\frac{I_n'(\theta)}{I_n(\theta)} = \frac{M_{n-1}'(\theta) + I_{n-1}(\theta)}{M_{n-1}(\theta)} . \tag{22}$$

Combining (21) and (22), we see that

$$\frac{I_n'(\theta)}{I_n(\theta)} = \frac{I_{n-1}'(\theta)}{I_{n-1}(\theta)} = \cdots = \frac{I_1'(\theta)}{I_1(\theta)} = \kappa(\theta) ,$$

from which the desired result follows upon noting that $I_n(\theta) \in \mathcal{F}_{n-1}$.

Conversely, if

$$M_n(\theta) = \phi(\theta) H_n(\hat{\theta}_n - \theta)$$

for some $\mathcal{F}_{n-1}$-measurable random variable H_n and any $n \geq 1$, notice first that

$$u_n(\theta) = \phi(\theta)(H_n(\hat{\theta}_n - \theta) - H_{n-1}(\hat{\theta}_{n-1} - \theta)) ,$$

where $H_0 = 0$. Differentiating and computing conditional expectations yields

$$\begin{aligned} \mathbf{E}(\nu_n(\theta) | \mathcal{F}_{n-1}) = &- \phi'(\theta) H_n \mathbf{E}(\hat{\theta}_n - \theta | \mathcal{F}_{n-1}) + \phi(\theta) H_n + \phi'(\theta) H_{n-1}(\hat{\theta}_{n-1} - \theta) \\ &- \phi(\theta) H_{n-1} . \end{aligned}$$

But $\mathbf{E}(u_n(\theta)|\mathscr{F}_{n-1}) = 0$, so the previous equation simplifies to

$$\mathbf{E}(\nu_n(\theta)|\mathscr{F}_{n-1}) = -\phi(\theta)(H_n - H_{n-1}),$$

whence

$$\mathbf{E}u_i(\theta)^2 = \phi(\theta)(H_i - H_{i-1}), \tag{23}$$

and

$$I_n(\theta) = \sum_{i=1}^{n} \mathbf{E}(u_i(\theta)^2|\mathscr{F}_{i-1}) = \phi(\theta)H_n. \qquad \square$$

Corollary: Under the model (19), $(H_n, \hat{\theta}_n)$ is a minimal sufficient statistic.

Proof: Sufficiency is immediate from the factorization criterion and the proposition. To see the minimality, let $Z_1, \ldots, Z_n$ and $Z_1^*, \ldots, Z_n^*$ be two paths with corresponding likelihoods L_n and L_n^*. Since $L_n(\theta) = H_n(\Phi(\theta)(\hat{\theta}_n - \Psi(\theta))$, where $\Phi'(\theta) = \phi(\theta)$ and $\Psi'(\theta) = \theta\phi(\theta)$ we have that

$$\frac{L_n(\theta)}{L_n^*(\theta)} = \exp\{H_n(\Phi(\theta)\hat{\theta} - \Psi(\theta)) - H_n^*(\Phi(\theta)\hat{\theta}^* - \Psi(\theta))\}$$

which is independent of θ if and only if $\hat{\theta} = \hat{\theta}^*$ and $H_n = H_n^*$. $\square$

Notice that the maximum likelihood estimator in the model (19) is not by itself a sufficient statistic (except when H_n is constant), in contrast to the corresponding iid model (where H_n is constant). One may explain the generally good performance of mles in the iid case by the fact that they are *asymptotically sufficient* (meaning that the likelihood factors approximately for large n) under mild regularity conditions (Cox and Hinkley, 1974, section 9.2(v)), even when there is no exactly sufficient statistic. In the general nonergodic case (where (15) holds but not necessarily (19)), sufficiency may not even occur asymptotically. Write

$$\log L_n(\theta) = \log L_n(\hat{\theta}_n) + \tfrac{1}{2}(\hat{\theta}_n - \theta)^2 M_n'(\hat{\theta}_n) + o_{\mathbf{P}}(1).$$

Using (15) we see that $M_n'(\theta)/I_n(\theta) \xrightarrow{\mathbf{P}} -1$ as $n \to \infty$. Thus

$$L_n(\theta) = L_n(\hat{\theta}_n)\exp(-\tfrac{1}{2}H_n\phi(\theta)(\hat{\theta}_n - \theta)^2 + o_{\mathbf{P}}(1)). \tag{24}$$

Even though $\hat{\theta}_n$ is asymptotically efficient, it is not asymptotically sufficient, since the factor multiplying $L_n(\hat{\theta}_n)$ depends both on $\hat{\theta}_n$ and H_n. However, $\hat{\theta}_n$ cannot be improved by conditioning on the sufficient statistic, since it is already part of it. We will later argue that I_n may be thought of as asymptotically ancillary, so that inferences may be made conditional upon its value.

Before we apply this theory to BGW processes, we will first specialize to the Markov situation, where the distributions satisfying (19) take a particularly simple form. Since

$$L_n(\theta) = \prod_{i=1}^{n} f(X_i|X_{i-1}; \theta),$$

where $f(x|y; \theta)$ is the transition density, we have that

$$u_i(\theta) = \frac{\partial}{\partial\theta} \log f(X_i|X_{i-1}; \theta).$$

Letting $n = 1$ we see that

$$\frac{\partial}{\partial\theta} \log f(X_1|X_0; \theta) = \phi(\theta) H_1(\hat{\theta}_1 - \theta).$$

$\hat{\theta}_1$ is the unique solution $\mu(X_1, X_0)$ of the equation

$$\frac{\partial}{\partial\theta} f(X_1|X_0; \theta) = 0.$$

In other words, we can write the transition density

$$\frac{\partial}{\partial\theta} \log f(x|y; \theta) = \phi(\theta) H(y)(\mu(x, y) - \theta). \tag{25}$$

Such a transition density is said to be a member of a *conditional exponential family*. It is easy to see that the conditional exponential form is both necessary and sufficient for a Markov chain to satisfy (19).

Finally specializing to the BGW process setup, we first identify the conditional exponential family. The mle based on one observation is $\hat{m}_1 = Z_1/Z_0$, so $\mu(x, y) = x/y$ for integers $x \geq 0$ and $y \geq 1$. Consequently

$$u_i(m) = \phi(m) H(Z_{i-1}) \frac{Z_i - mZ_{i-1}}{Z_{i-1}},$$

so that

$$\mathbf{E}(u_i^2|\mathcal{F}_{i-1}) = \phi(m)^2 \sigma^2 H(Z_{i-1})^2 / Z_{i-1}.$$

From (23) and the representation $H_k = \Sigma_{i=1}^{k} H(Z_{i-1})$ we see that

$$\mathbf{E}(u_i(m)^2|\mathcal{F}_{i-1}) = \phi(m) H(Z_{i-1}),$$

so that

$$H(Z_{i-1}) = \frac{Z_{i-1}}{\sigma^2 \phi(m)}$$

and

$$\frac{\partial}{\partial m} \log p(x|y; m) = \frac{x - my}{\sigma^2} .$$

Letting $y = 1$ we see that p must be a power series distribution:

$$p_j = a_j \lambda^j A(\lambda)^{-1} .$$

Here (see Exercise 6) $m = \lambda(d/d\lambda) \log A(\lambda)$ and $\sigma^2 = ((d/dm) \log \lambda)^{-1}$. This distribution is the one-parameter exponential family for discrete data. The likelihood for this model is

$$L_n(\lambda) \propto \exp(Y_n \log \lambda - Y_{n-1} \log A(\lambda)) .$$

This is the curved subset $\{\log \lambda, -\log A(\lambda)\}$ of the natural parameter space of the full exponential family with density $\exp(Y_n \eta_1 - Y_{n-1} \eta_2 - K(\eta_1, \eta_2))$. We will return to this in sections 9 and 10. The computations in the Examples of section 7 show that $I_n(m) = Y_{n-1}/\sigma^2$. You may want to check that (19) holds. Using the Corollary to Theorem 1.1 we see that

$$\frac{I_n(m)}{\xi_n(m)} \to W \text{ a.s.}$$

and it is clear that $\hat{m}_n$ is Rao-efficient on A when $m > 1$. Using the same normalizing factor we have that

$$I_n(m)^{1/2} \left(\frac{m-1}{m} \right)^{1/2} (\bar{m}_n - m) \xrightarrow{d} \mathbf{N}(0, 1) ,$$

so that the normalizing factor in (14) is

$$\gamma(m) = \left(\frac{m}{m-1} \right)^{1/2} ,$$

meaning that the asymptotic relative efficiency of $\bar{m}$ relative to $\hat{m}$ is $1 - 1/m$. Since m_n^* converges at the rate n, rather than at the rate $I_n = O(m^n)$, it has asymptotic relative efficiency 0. We have shown the following.

Theorem 2.7: Conditional upon nonextinction, the asymptotic relative efficiency of $\bar{m}_n$ to $\hat{m}_n$ for the power series offspring distribution is $1 - 1/m$, while m_n^* is not Rao-efficient.

It would be preferable to have an optimality concept that is not asymptotic, and does not need to be interpreted conditionally. Godambe (1960, 1985) proposed a finite sample optimality property for estimators, which is based on ideas very similar to those underlying Rao-efficiency. Let

$(X_k, k = 1, \ldots, n)$ be a stochastic process, and let $T = T(X_1, \ldots, X_n; \theta)$ be a function satisfying

$$\mathbf{E}_\theta T(X_1, \ldots, X_n; \theta) = 0\,.$$

Any such function is called an unbiased estimating function, and the equation $T = 0$, provided that it has a solution in Θ, provides an estimate of θ. The concept of unbiased estimating functions is a generalization of Fisher consistency to more general estimators. The optimum estimating function g^* within the class of unbiased estimating functions is that satisfying

$$\mathbf{E}_\theta(g^* - \eta(\theta)M_n(\theta))^2 \le \mathbf{E}_\theta(g - \eta(\theta)M_n(\theta))^2$$

for all $\theta \in \Theta$ and all unbiased estimating functions g. Under mild regularity conditions, g^* is optimal if it is most highly correlated with the underlying score function:

$$\mathbf{Corr}_\theta(g^*, M_n(\theta)) \ge \mathbf{Corr}_\theta(g, M_n(\theta))\,.$$

In order to exhibit an optimal estimating function, one may wish to restrict attention to the class of linear estimating functions

$$g = \sum_1^n h_i a_{i-1}\,,$$

where $\mathbf{E}_\theta(h_i(\theta)|\mathcal{F}_{i-1}^X) = 0$ and a_{i-1} may depend on θ and $x_1, \ldots, x_{i-1}$. In this class the optimal estimating function is given by choosing

$$a_{i-1}^* = \frac{\mathbf{E}(\partial h_i/\partial\theta|\mathcal{F}_{i-1}^X)}{\mathbf{E}(h_i^2|\mathcal{F}_{i-1}^X)}\,.$$

Example: Branching Process

Suppose we want to estimate the offspring mean of a BGW branching process. If we put $h_i = Z_i - mZ_{i-1}$ we have that

$$\mathbf{E}(h_i|\mathcal{F}_{i-1}^Z) = 0$$

and

$$\mathbf{E}(h_i^2|\mathcal{F}_{i-1}^Z) = Z_{i-1}\sigma^2\,.$$

Hence the optimal estimating equation is

$$g^* = \sum_{i=1}^n (z_i - mz_{i-1}) = 0\,.$$

The solution is $\hat{m}$. We see that $\bar{m}$ satisfies

$$g = z_n / z_{n-1} - m = 0$$

and therefore is not optimal within this class of estimating functions.

Suppose now instead that we wish to estimate the variance, for simplicity assuming that the mean is known. Let

$$h_i = (Z_i - mZ_{i-1})^2 - \sigma^2 Z_{i-1} .$$

Then

$$\mathbf{E}(\partial h_i / \partial\theta | \mathcal{F}_{i-1}^Z) = -Z_{i-1}$$

and

$$\mathbf{E}(h_i^2 | \mathcal{F}_{i-1}^Z) = \sigma^4((k-3)Z_{i-1} - Z_{i-1}^2) ,$$

where k is the kurtosis (the standardized fourth central moment) of the offspring distribution. The optimal estimating equation has solution

$$\sigma^{*2} = \frac{\sum \dfrac{(Z_i - mZ_{i-1})^2}{k-3+2Z_{i-1}}}{\sum \dfrac{Z_{i-1}}{k-3+2Z_{i-1}}} .$$

We see that this depends crucially on the value of k, which we do not know (and which is difficult to estimate). In the case $k = 3$ we obtain an estimator due to Dion (1975; see section 3.1). Most standard discrete distribution (except the uniform) have kurtosis larger than 3, with the excess usually decreasing with m. If Z_{i-1} becomes large, the influence of k is relatively small, and Dion's estimator is approximately optimal in the sense discussed here. On the other hand, on a path that becomes extinct or nearly so, the value of k is quite important to the estimated value of σ^2. Hence this concept of optimality may not be all that useful for practical applications of branching processes. We shall discuss the estimation of σ^2 further in chapter 3. □

8. EXTREME POINT MODELS

The basis for the conditional statements made in the previous two sections has been a certain mathematical convenience. From a theoretical point of view this is not very satisfying: one would prefer to make these conditional statements on the basis of a statistical principle which necessitates these statements. In this and the following sections we will discuss some possible such principles.

Lauritzen (1984) surveys the general theory of *extreme point models* in statistics. He argues that statistical modelers should attempt to choose such models whenever possible. Many cases of conditional inference can in fact be interpreted as a modification of the model. To illustrate the ideas involved, we will first look at a simple example.

Example: de Finetti's Theorem

Let $X_1, X_2, \ldots$ be $\{0, 1\}$-valued random variables, and assume that they are *exchangeable*, i.e., that

$$(X_1, \ldots, X_n) \stackrel{d}{=} (X_{\pi(1)}, \ldots, X_{\pi(n)})$$

for all $n \in \mathbb{N}$ and all permutations π of the integers $\{1, \ldots, n\}$. The class of exchangeable random variables contains the independent Bernoulli sequences. de Finetti (1931) proved that there is a unique probability measure μ on $[0, 1]$ such that for all $n \in \mathbb{N}$

$$\mathbf{P}\{(X_1, \ldots, X_n) = (x_1, \ldots, x_n)\} = \int_0^1 \theta^{\Sigma x_i}(1-\theta)^{n-\Sigma x_i} \mu\,(d\theta)\,.$$

Moreover, the average $\bar{X} = \Sigma X_i/n$ converges a.s. to a random variable X_∞, and μ is the distribution of X_∞. Some reflection will show that the exchangeability condition holds if and only if for any $t \in \{0, \ldots, n\}$

$$\mathbf{P}\{(X_1, \ldots, X_n) = (x_1, \ldots, x_n) | X_1 + \cdots + X_n = t\} = \binom{n}{t}^{-1} 1\left(\sum x_i = t\right).$$

In other words, $T_n(X) = \Sigma X_i$ must be sufficient, with the conditional distribution of the data given T_n being uniform. If we look at the class $\mathbf{C}$ of all random variables on $\{0, 1\}^{\mathbb{N}}$ having the property that the sum is a sufficient statistic, and that the conditional distribution of the data given the sum is uniform, this is a *convex class*: if μ_1 and μ_2 are in $\mathbf{C}$, then $\lambda\mu_1 + (1-\lambda)\mu_2 \in \mathbf{C}$ for any $\lambda \in [0, 1]$. Furthermore, $\mathbf{C}$ is exactly the class of exchangeable binary random variables. An interesting subclass of any convex set $\mathbf{C}$ is the set $\mathscr{E}(\mathbf{C})$ of *extreme points*. We call $\mu \in \mathbf{C}$ an extreme point if $\mu = \lambda\mu_1 + (1-\lambda)\mu_2$ for some $\lambda \in (0, 1)$ implies that $\mu_1 = \mu_2$; in other words, $\mu \in \mathscr{E}(\mathbf{C})$ if it cannot be written as a linear combination of other members of $\mathbf{C}$. By de Finetti's representation, all exchangeable random variables are mixtures of binomial random variables, with the mixing distribution μ. Therefore they are linear combinations of binomials, except when μ concentrates on a single point. Thus in this case $\mathscr{E}(\mathbf{C})$ is precisely the set of independent Bernoulli random variables with fixed success probability.

From a statistical point of view, we would typically be interested in determining the distribution $\mathbf{P}$ of the data. Since the sufficient statistic contains all the available information about $\mathbf{P}$, and the limiting distribution of the sufficient statistic is μ, we can only identify $\mathbf{P}$ from the data if μ has a one-point distribution (since we even in the limit only observe a single value of the sufficient statistic), i.e., if our model is an extreme point model. The added generality of the class of exchangeable random variables is obtained at the price of not being able to consistently estimate the underlying distribution, even from an infinite stretch of observations. □

The Bernoulli case is the model for the general definition of extreme families. The discussion given here will be couched in substantially less general terms than in Lauritzen's work (see also Lauritzen, 1982, 1988). Suppose that we have a sequence of $X_1, X_2, \ldots$ of random variables with distribution absolutely continuous with respect to some measure μ_θ, indexed by $\theta \in \Theta$. Also consider a sequence of statistics $Y_i = T_i(\mathbf{X}_i)$, $i = 1, 2, \ldots,$ where $\mathbf{X}_i = (X_1, \ldots, X_i)$, such that

(i) the conditional distribution of $\mathbf{X}_i$ given $Y_i = y_i$ is uniform on $\{\mathbf{x}_i\colon T_i(\mathbf{x}_i) = y_i\}$,
(ii) for $j > i$, $\mathbf{X}_i$ and Y_j are conditionally independent given Y_i.

Notice that condition (i) implies that (Y_i) is a sufficient vector of statistics, provided that $\{\mathbf{x}_i\colon T_i(\mathbf{x}_i) = y_i\}$ does not depend on θ. Condition (ii) ensures a consistent system of probability distributions. The *maximal* family $\mathbf{M}$ corresponding to these statistics and conditional distributions is the set of all probabilities for which a de Finetti-type decomposition

$$\mathbf{P}(\mathbf{X}_i \in B) = \int \mathbf{P}(\mathbf{X}_i \in B \mid Y_i = y_i)\, d\mathbf{P}(Y_i = y_i)$$

holds for each i. This is the largest family such that the Y_i are sufficient statistics for any family of the form $\{\mu_\theta, \theta \in \Theta) \subseteq \mathbf{M}$. $\mathbf{M}$ is a convex set, and the set $\mathscr{E}(\mathbf{M})$ of extreme points of $\mathbf{M}$ is called the *extremal family* corresponding to the given sequence of statistics and conditional distributions.

The class $\mathbf{M}$ is a *simplex*, in the sense that for any measure $\mu \in \mathbf{M}$ there is precisely one probability $\mathbf{P}_\mu$ defined on $\mathscr{E}(\mathbf{M})$, such that

$$\mu(A) = \int_{\mathscr{E}(\mathbf{M})} e(A)\mathbf{P}_\mu\,(de)\,.$$

It turns out that the extreme point models are precisely those for which the tail sigma-field $\mathscr{T} = \bigcap_i \sigma(Y_j,\ j > i)$ is trivial (meaning that all tail events have probability zero or one). Define for fixed i and A

$$Z_n(A) = \mu(\mathbf{X}_i \in A \mid Y_n)\,, \qquad n > i\,.$$

Then $Z_n(A)$ is a bounded reverse martingale, with expectation $\mu(A)$, and it converges a.s. to $Z_\infty(A) = \mu(A|\mathcal{T})$. In other words, $Z_n(A)$ is an estimator of $\mu(A)$. It is consistent precisely when $\mu \in \mathscr{E}(\mathbf{M})$. Of course, from a single realization of the process we cannot tell whether or not Z_∞ is random. Lauritzen uses this observation as basis for a recommendation always to choose an extreme model: it permits consistent estimation from a single realization, and if observations support a nonextreme model, they also support an extreme model.

Example: Poisson Distribution

Let $X_1, X_2, \ldots$ be independent Poisson variates with $\mathbf{E}_\theta X_n = \theta^n$ for some $\theta > 0$. The likelihood is

$$\left(\prod x_i!\right)^{-1} \theta^{\Sigma\, ix_i} e^{-\Sigma\, \theta^i} .$$

so we have an exponential family with sufficient statistic

$$T_n(x_1, \ldots, x_n) = \sum_{i=1}^{n} ix_i .$$

The conditional distribution of the observations, given the statistic, is

$$q(x_1, \ldots, x_n | T_n = t) = c_n(t)^{-1} \qquad \text{if} \sum ix_i = t ,$$

where

$$c_n(t) = \sum_{k=0}^{\infty} \frac{1}{k!} \sum_{\Sigma\, ix_i = 1} \binom{k}{x_1, \ldots, x_n} .$$

Note that if $\theta < 1$ we have

$$\sum \mathbf{E}|iX_i| = \sum i\theta^i < \infty$$

and

$$\mathbf{Var} \sum_{1}^{\infty} iX_i = \sum i^2\theta^i < \infty ,$$

so $Y_\infty = \Sigma_{i=1}^{\infty} iX_i$ is a tail measurable random variable with nondegenerate distribution. Hence $\mathbf{P}_\theta$ is not extreme for $0 < \theta < 1$. One can show (Lauritzen, 1982, pp. 119–125) that $\mathbf{P}_\theta$ is extreme if and only if $\theta \geq 1$. But these are not the only extreme points: the distributions obtained by conditioning on Y_∞ for some $\theta < 1$ are also extreme. Thus if $\theta < 1$ conditioning on the tail-measurable random variable Y_∞ is the appropriate thing to do. Put differently, if $\theta < 1$ the experiment performed yields consistent information only about Y_∞, not about θ itself. □

Example: Positive Geometric Distribution

Now consider a BGW process with positive geometric offspring distribution. Then $m^{-n}Z_n$ converges a.s. to a standard exponential random variable W, and conditional upon the value w of W, $Z_n - Z_{n-1}$ are independent, Poisson distributed random variables with geometrically increasing expectations $w(m-1)m^{n-1}$ (Athreya and Ney, 1972, Theorem III.11.2 and Harris, 1963, chapter V, sections 5 and 8). Hence the extreme points of this *conditional* model are the same as those in the previous example. In fact, Lootgieter (1977) showed that $\mathscr{T} = \sigma(W)$. Keiding (1974) suggested similar inference for birth processes. □

9. ANCILLARITY

The corollary to Proposition 2.12 shows that $(H_n, \hat{\theta}_n)$ is a sufficient statistic in the exponential family model (19). Thus, a two-dimensional statistic is sufficient for a one-dimensional parameter. In the branching process case, using a power series offspring distribution, the sufficient statistic is $(Y_{n-1}, \hat{m}_n)$. The results of section 7 show that $\hat{m}_n$ is an efficient statistic, even though it does not carry all the information in the sample about m. Thus it would appear that Y_{n-1} somehow asymptotically does not carry very much information about the value of m. To make this a little more precise, we can use Proposition 2.7 to see that on the explosion set Y_{n-1} is asymptotically normally distributed. The mean and variance are derived in Exercise 1.2(b), and are seen to be

$$\mathbf{E}Y_{n-1} = \frac{m^n - 1}{m - 1}$$

and

$$\mathbf{Var}\ Y_{n-1} = \frac{\sigma^2}{(m-1)^2}\left(\frac{m^{2n-1} - 1}{m-1} - (2n-1)m^{n-1}\right).$$

For large n it is straightforward to compute the Fisher information in the limiting distribution, which by dominated convergence is asymptotically correct. Ignoring terms of order m^{-n} or smaller, we get

$$\xi_{Y_{n-1}}(m) = O\left(n^2\left(\frac{4}{m^2} + \frac{1}{\sigma^2}\right)\right).$$

Since the Fisher information in the observed process is $\mathbf{E}Y_{n-1}/\sigma^2$, we see that the ratio of information is $O(n^2 m^{-n})$ which converges to 0 as $n \to \infty$. Hence, in this sense, there is only negligible information about m in Y_{n-1}. Theorem 2.3 indicates that Y_{n-1} plays the role of the sample size in the iid setup, indicating the precision of $\hat{m}_n$. This is a quantity that Fisher (1973) terms an ancillary statistic, explaining (p. 165):

> ... loss of information by simple estimation is due to the variance in the amount of information "realized". A simple remedy, appropriate for this asymptotic situation is merely to record, not only the estimate of maximum likelihood but the amount of information realized, or the "apparent precision", using this to supply a "weight" for the estimate more precise than the supplied by the size of the sample.

There has been several different attempts to make this argument more precise. The simplest definition of an ancillary statistic arises when the sufficient statistic is $(T_n, \hat{\theta}_n)$, where the distribution of T_n does not depend on θ. It is then natural to carry out inference conditional on the value of T_n, and no information about θ is lost, in the sense that the Fisher information in the joint distribution of $(T_n, \hat{\theta}_n)$ is the same as in the conditional distribution of $\hat{\theta}_n$, given T_n. In cases where the distribution of T_n is not free of θ, one may look for some function of T_n that has an asymptotic distribution that is θ-free, and if that fails, some function of T_n that has a distribution that does not depend on θ very much, e.g., having mean and variance free of θ. This is the route advocated by Efron and Hinkley (1978), and in a nonergodic stochastic process context by Feigin and Reiser (1979). For an approximate ancillary that is asymptotically normally distributed, this would typically imply that asymptotically the Fisher information in T_n is negligible.

Example: The Positive Geometric Distribution

In the case of a positive geometric offspring distribution, we know that W has an exponential distribution, free of m. Therefore we can argue that we may want to make our inferences conditional upon $W_n = Y_{n-1}/\mathbf{E}Y_{n-1}$, which is asymptotically ancillary. Equivalently (from the point of view of probability theory) we may condition on Y_{n-1}. □

The event A of eventual explosion is, as we have noted before, a tail measurable event. In nonergodic processes the tail σ-field $\mathcal{T}$ is nontrivial. As pointed out in the previous section, $\mathcal{T} = \sigma(W)$, where W is the limiting random variable from Theorem 1.1. By conditioning on W the model is reduced to an ergodic one. This amounts to regarding W as a nuisance parameter. However, since W has a known frequency distribution, it would seem reasonable to integrate it out, rather than condition on it. Conditioning, however, is the correct procedure if we regard the unconditional model as a mixture model, with W the mixing variable. In addition, as seen in the previous section, conditioning may yield an extreme model, if one is to be had. The conditional distribution given W is hard to compute in general, and the actual value of W is unknown, although $\hat{m}^{-n}Z_n$ is a good approximation for large n. Similarly, it is difficult to marginalize over the distribution of W.

Example: The Positive Geometric Distribution

One of the few cases in which we can compute the conditional distribution given W is the positive geometric case (see the last Example of the previous section). If we consider conditional inference given W appropriate, the likelihood becomes, with $\theta_n = w(m-1)m^{n-1}$ and $\delta_n = Z_n - Z_{n-1}$

$$L_n(m) \propto e^{-\Sigma \theta_n} \prod_1^n \theta_n^{\delta_n},$$

and the conditional mle solves the equation

$$\sum \delta_n \left(\frac{1}{m-1} + \frac{n-1}{m} \right) = w \sum \left(1 + \frac{(m-1)(n-1)}{m} \right) m^{n-1},$$

which, of course, depends on w. □

A much weaker requirement than extremity for asymptotic statistical inference is the following.

Tail Conditioning Principle: Asymptotic inference for nonergodic models should be made conditional upon an appropriate tail event.

There is considerable ambiguity in this principle, although it seems to be implicit in most work on asymptotic methods for nonergodic models. For example, can we use any tail event? From a finite path, how do we decide which tail event we are in? In the literature on inference for branching processes this principle is usually adhered to, in that inference is made conditional upon explosion, which of course is a tail event.

Since Theorem 2.3 asserts mixing convergence, we can deduce that

$$\left(\sigma(m)^{-1} Y_{n-1}^{1/2}(\hat{m}_n - m), \frac{Y_{n-1}}{\mathbf{E}Y_{n-1}} \right) \xrightarrow{d} (Z, W), \tag{26}$$

where $Z \sim \mathbf{N}(0,1)$ and W is independent of Z. Denote $W_n = Y_{n-1}/\mathbf{E}Y_{n-1}$, and $X_n = Y_{n-1}^{1/2}(\hat{m}_n - m)$. We define $\gamma_n^2(m) = (1 - \rho_n^2(m))\, \mathbf{Var}\, W_n$, where ρ_n is the correlation coefficient between W_n and $\Delta_n = W_n^{1/2} X_n$. The latter quantity is the *score statistic*, i.e., the score martingale normalized by the square root of the expected information. Typically, $\rho_n \to 0$, so $\gamma_n^2 \to \mathbf{Var}\, W = \sigma^2/m(m-1)$ (see Theorem 1.1(iv)). The quantity γ_n is called the *statistical curvature* of the problem. It was introduced under that name by Efron (1975), and was shown to measure how fast a statistical problem is changing, in the sense of power of a local test at statistically interesting (local) alternatives (see section 10). Families with low curvature are like exponential families, where the locally most powerful test of a one-sided hypothesis is uniformly most powerful, whereas for those with high curvature the locally most powerful test behaves poorly. Also, in nonexponential families, the mle is not generally sufficient. The loss of information in using the mle can rather than the full sufficient statistic be expressed in terms of the statistical curature, and corresponds to $\gamma_n(\theta)^2$ observations. More formally, let

$$\sum_\theta = \begin{pmatrix} \mathbf{E}_\theta \dot{l}_\theta^2 & \mathbf{E}_\theta \dot{l}_\theta \ddot{l}_\theta \\ \mathbf{E}_\theta \ddot{l}_\theta \dot{l}_\theta & \mathbf{E}_\theta \ddot{l}_\theta^2 - i_\theta^2 \end{pmatrix}$$

where i_θ is the Fisher information, **Var** $\dot{l}_\theta$, l_θ is the log likelihood, and we suppress the dependence on n. Then the curvature is given by

$$\gamma(\theta) = \left(\det \sum_\theta \Big/ i_\theta^3\right)^{1/2} .$$

In our situation we can compute the curvature of the BGW branching processes with power series offspring distribution, namely

$$\gamma_n^2(m) = m^{n-1} \frac{(m^n - 1)^2 - n^2 m^{n-1}(m-1)^2}{(m^n - 1)^3 (m-1)} \sigma^2 .$$

We see that for $m > 1$ the leading term in γ_n^2 is $\sigma^2/m(m-1) = \mathbf{Var}\, W$. It should be pointed out that in an ergodic setup the curvature converges to zero as $n \to \infty$, whereas in the nonergodic situation it will converge to a nonzero function of the parameter of interest.

Example: The Positive Geometric Distribution

In the positive geometric case we have that

$$\gamma_n^2 = \frac{1}{1 - m^{-n}} - \frac{n^2 (m-1)^2}{m^{n+1}(1 - m^{-n})^3}$$

is asymptotic to 1, in accordance with the ancillarity of W. The convergence is nearly exponentially fast, since $(m^n/n^2)(1 - \gamma_n^2) = (m-1)^2/m + o(1)$.

This is essentially the only one-parameter conditional exponential family model with constant curvature. To see that, note that the curvature will converge to a constant whenever $\sigma^2 = cm(m-1)$. In the class of power series distributions, where $\sigma^2 = ((d/dm) \log \theta)^{-1}$, this condition implies that $\theta = (1 - 1/m)^c$, so $m = (1 - \theta)^{-1/c}$, whence $A(\theta)$ satisfies

$$\frac{d}{d\theta} \log A(\theta) = (\theta(1 - \theta^{1/c}))^{-1}$$

so

$$A(\theta) = (1 - \theta^{1/c})^{-1} ,$$

corresponding to a geometric distribution with parameter $\theta^{1/c}$. □

Now introduce

$$Q_n(m) = \frac{W_n - 1}{\gamma_n(m)} .$$

Since $W_n \to W$ with $\mathbf{E}W = 1$, and $\gamma_n \to \mathbf{Var}\, W$, $Q_n(m)$ behaves asymptotically like a standardized version of W. Efron and Hinkley (1978) introduced $Q_n(\hat{m})$ as an ancillary statistic in the vein of Fisher's discussion above. In the iid setup, Q_n has asymptotically a standard normal distribution, so it is asymptotically an ancillary statistic in the sense of having a distribution free of the parameter under study. Feigin and Reiser (1979) noted that in the nonergodic situation Q_n is only approximately asymptotically ancillary: higher moments will depend on m. For known m, $Q_n(m)$ is a linear function of Y_{n-1}, lending some credence to an assertion that Y_{n-1} is an appropriate approximate ancillary (notice the two approximations: one for large n, another for higher moments) for the problem of estimating m. Inference then should be carried out using the conditional distribution of $\hat{\theta}_n$ given the value of Q_n (or W_n). However, if this conditional distribution ought to be used, it may be argued that one should also compute a conditional maximum likelihood estimate instead of $\hat{\theta}_n$.

Sweeting (1986) argues that Y_{n-1} can be considered ancillary, since its asymptotic probability mass function is continuous in m, and therefore is approximately constant over the range of $\hat{m}_n$. He defines asymptotic ancillarity in a local sense by

$$\lim_{n\to\infty} \frac{\mathbf{P}(Y_{n-1} = y_n | m_n)}{\mathbf{P}(Y_{n-1} = y_n | m)} = 1 , \tag{27}$$

whenever $|m_n - m| \le Am^{-1/2n}$ and $y_n / c_n \to w$ for any $A > 0$, $w > 0$, where $c_n = \mathbf{E}Y_{n-1} = (m^n - 1)/(m - 1)$.

Theorem 2.8: Assume that $p_0 = 0$. Then Y_{n-1} is asymptotically ancillary for m in the sense of (27). Furthermore, the conditional distribution of X_n, given Y_{n-1}, converges in distribution to $\mathbf{N}(0, \sigma^2)$, uniformly in compact intervals of $m \in (1, \infty)$.

Sketch of Proof: We first derive a formula for the conditional characteristic function. Let (ξ, η) be random variables with the same lattice size l and joint characteristic function $\phi(s, t) = \mathbf{E}\exp(i(s\xi + t\eta))$. Write the conditional characteristic function of ξ given that $\eta = y$

$$\phi(s|y) = \sum_x \exp(isx)p(x|y)$$

so that

$$\phi(s, t) = \sum_y \exp(ity)p(y)\phi(s|y) .$$

Here $p(x|y)$ is the conditional density of ξ given η, and $p(y)$ is the marginal density of η. By Fourier inversion

$$p(y)\phi(s|y) = \frac{l}{2\pi}\int_{-2\pi/l}^{2\pi/l} \exp(-ity)\phi(s,t)\,dt$$

and

$$p(y) = p(y)\phi(0|y) = \frac{l}{2\pi}\int_{-2\pi/l}^{2\pi/l} \exp(-ity)\phi(0,t)\,dt$$

Writing

$$p^s(y) = \frac{l}{2\pi}\int_{-2\pi/l}^{2\pi/l} \exp(-ity)\phi(s,t)\,dt\,,$$

and noting that $p^0(y)$ is just the marginal density $p(y)$, we see that

$$\phi(s|y) = \frac{p^s(y)}{p(y)}\,. \tag{28}$$

Applying this formula to the branching process situation, let ξ be $U_n = W_n^{1/2}X_n$ and η be Y_{n-1}. Then $\phi_n(s|y) = p_n^s(y)/p_n(y)$. The key point of the proof is to verify that

$$\frac{1}{2\pi}\int_{-\pi c_n/l}^{\pi c_n/l} \exp(-iwt\phi_n(s,t)\,dt \to \phi(s|w)f(w) \tag{29}$$

holds locally uniformly in $s \in \mathbf{R}$, uniformly for m in compact intervals of $(1,\infty)$, where $f(w)$ is the density of W, $\phi(s|w) = \mathbf{E}\exp(isW^{1/2}Z|W=w)$, and Z is an $\mathbf{N}(0,\sigma^2)$ random variable, independent of W. The proof of this result hinges on careful analysis of the joint characteristic function of (X_n, W_n), and is not given here. To use the result, let (y_n) be a sequence of possible values of Y_{n-1}, such that $y_n/c_n \to w > 0$. By (28), $\phi_n(s|y_n)p_n(y_n) = p_n^s(y_n)$, and by (29), $\phi_n(s|y_n) \to \phi(s|w)$ locally uniformly in s. Thus for each $m > 1$

$$\begin{aligned}\mathbf{E}(\exp(isX_n|Y_{n-1}=y_n)) &= \mathbf{E}(\exp(is(y_n/c_n)^{-1/2}U_n)|Y_{n-1}=y_n)\\ &\to \mathbf{E}(\exp(isw^{-1/2}U)|W=w) = \mathbf{E}\exp(isZ)\,.\end{aligned}$$

Careful analysis of the proof of (28) verifies that the convergence is uniform for m in compact subsets of $(1,\infty)$. Finally, let $s=0$ to see that

$$\frac{\mathbf{P}(Y_{n-1}=y_n|m_n)}{\mathbf{P}(Y_{n-1}=y_n|m)} = \frac{l^{-1}c_n(m_n)p_n(y_n;m_n)}{l^{-1}c_n(m)p_n(y_n;m)}\,\frac{c_n(m)}{c_n(m_n)}$$

which converges to 1 provided that the ratio $c_n(m)/c_n(m_n)$ converges to 1 as $n \to \infty$, which is the case if and only if $m_n - m = o(1/n)$. □

Notice that convergence of the joint distribution of two random variables need not imply convergence of the conditional distribution of one given the other, so this result is stronger than what we have seen before. The result requires that $p_0 = 0$, but a similar result conditional upon nonextinction can presumably be proved. Under suitable regularity conditions local asymptotic ancillarity is equivalent to having a vanishing fraction of Fisher information in the locally ancillary statistic. Sweeting's result proves a claim in Feigin and Reiser (1979) that asymptotic inference conditional upon Q_n is equivalent to using the asymptotic distribution of $I_n(\hat{m}_n)^{1/2}(\hat{m}_n - m)$.

In summary, there are several ways in which inference conditional upon Y_{n-1} can be motivated. On the other hand, the most common conditioning event is $\{Z_n > 0\}$. It is not an approximate ancillary in the sense of Feigin and Reiser or in the sense of statistical curvature. However, Sweeting (1986) points out that $\mathbf{P}(Z_n > 0|m) \to 1 - q(m)$ locally uniformly in m (this can be shown using the method suggested for solving Exercise 9), where the extinction probability $q(m)$ is a continuous function of m, and hence that conditioning upon $Z_n > 0$ can be motivated on grounds of local asymptotic ancillarity. This is the only statistical motivation advanced for such conditioning.

10. ASYMPTOTIC TESTING

Asymptotic likelihood methods for testing statistical hypotheses about nonergodic models behave quite differently from the iid case. The standard expansion of the log likelihood around a value θ^* close to θ is

$$l_n(\theta) = l_n(\theta^*) + (\theta - \theta^*)M_n(\theta^*) - \tfrac{1}{2}(\theta - \theta^*)J_n(\theta^*) + \text{error}\,. \qquad (30)$$

Let $W_n(\theta) = J_n(\theta)/\mathbf{E}_\theta J_n(\theta)$, so that $W_n(\theta) \xrightarrow{\mathbf{P}} W(\theta)$. When θ^* is taken to be $\hat{\theta}$, equation (24) follows in the form

$$L_n(\theta) = L_n(\hat{\theta}) \exp(-\tfrac{1}{2}(\hat{\theta} - \theta)^2 \mathbf{E}_\theta J_n(\theta) W_n(\theta))(1 + \text{error})\,.$$

When $W(\theta)$ is constant, i.e., in the ergodic case, this equation shows the asymptotic equivalence under the null hypothesis $\theta = \theta_0$ of the *likelihood ratio statistic*

$$\lambda_n = 2\Lambda_n(\hat{\theta}, \theta_0) = 2 \log L_n(\hat{\theta})/L_n(\theta_0)$$

and the Wald statistic (Cox and Hinkley, 1974, section 9.3, equation 41)

$$\omega_n = (\hat{\theta} - \theta_0)^2 \mathbf{E}_\theta J_n(\theta)|_{\theta = \hat{\theta}}\,.$$

This equivalence does not hold for the nonergodic case, since λ_n here is equivalent to $\omega_n W_n(\theta_0)$ under the usual continuity assumptions. Efron and

Hinkley (1978) argue that a better version of the Wald statistic (in finite samples) is

$$\omega_n^* = (\hat\theta - \theta)^2 J_n(\hat\theta) ,$$

which is seen to be asympotically equivalent to the likelihood ratio statistic also in the nonergodic case.

In the asymptotic theory of testing the statistically interesting alternatives θ_n are such that $\theta_n \to \theta_0$ as $n \to \infty$. This is because alternatives at a fixed distance can asymptotically be detected with probability one. In the branching process case we must further restrict our attention to supercritical processes only: since the asymptotics are different between supercritical and subcritical processes, it is not possible to test the perhaps most interesting hypothesis, namely that of distinguising between supercritical and subcritical processes. Let $\theta_n = \theta_0 + \xi_n^{1/2}(\theta_0)h$ where $\xi_n = \mathbf{E}_{\theta_0} J_n(\theta_0)$ and h is a fixed number. Then the log likelihood ratio is from (30)

$$\Lambda_n(\theta_n, \theta_0) = h\Delta_n(\theta_0) + \tfrac{1}{2}h^2 W_n(\theta_0) + o_{\mathbf{P}_\theta}(1) , \tag{31}$$

where $\Delta_n(\theta_0) = \xi_n(\theta_0)^{-1/2} M_n(\theta_0)$ is the *score statistic* for testing $\theta = \theta_0$ against $\theta > \theta_0$ (Cox and Hinkley, 1974, section 9.3, equation 42). Since Λ_n is the Neyman–Pearson test statistic for $\theta = \theta_0$ against $\theta = \theta_n$ with $h > 0$, it is clear that as $h \to 0$ this becomes equivalent to the score statistic, which therefore is locally most powerful (cf. the argument in Rao, 1973, section 7a.4). From (31) we also see that in the ergodic case, λ_n is asymptotically equivalent to Δ_n, while in the nonergodic case, they are no longer asymptotically equivalent.

Example: Power Series Offspring Distribution

From the computations in the Examples of section 7 we have that

$$M_n(m) = \sigma^{-2}(m)(Y_n - 1 - mY_{n-1}) ,$$

$$I_n(m) = \sigma^{-2}(m) Y_{n-1} ,$$

$$J_n(m) = -\frac{d}{dm}(\log \sigma^{-2}(m)) M_n(m) + \sigma^{-2}(m) Y_{n-1} ,$$

and

$$\xi_n(m) = \sigma^{-2}(m) \frac{m^n - 1}{m - 1} .$$

Hence the score statistic is

$$\Delta_n(m) = \frac{Y_n - 1 - mY_{n-1}}{\sigma(m)\left(\dfrac{m^n - 1}{m - 1}\right)^{1/2}} = \left(\frac{I_n}{\xi_n}\right)^{1/2} I_n^{1/2}(\hat m - m) ,$$

while the Wald statistic is

$$\omega_n(m) = \left(\frac{m^n - 1}{(m-1)\sigma^2(m)}\right)\left(\frac{Y_n - 1 - mY_{n-1}}{Y_{n-1}}\right)^2 = \left(\frac{\xi_n}{I_n}\right) I_n(\hat{m} - m)^2 \,,$$

and the likelihood ratio statistic

$$\lambda_n = 2(Y_n - 1)\log\frac{\hat{\lambda}}{\lambda} - 2Y_{n-1}\log\frac{A(\hat{\lambda})}{A(\lambda)}$$

which by Taylor expansion is asymptotically equivalent to

$$\sigma^{-2}(m)Y_{n-1}(\hat{m} - m) = I_n(\hat{m} - m)^2 \,.$$

Using slightly different (but asymptotically equivalent) normalizing factors, the Wald statistic corresponds to $W_n\lambda_n$ and the squared score statistic to $W_n^{-1}\lambda_n$. □

Let $\mathscr{A}$ be the class of all test function ϕ_n based on $\mathscr{F}_n^Z$, such that the power function $\pi_n(\theta; \phi_n) = \mathbf{E}_\theta\phi_n$ is continuous in θ, with $\alpha_n = \pi_n(\theta_0) \to \alpha$ as $n \to \infty$. In other words, $\mathscr{A}$ is the class of asymptotic size α tests with continuous power function. We say that $\phi_n^0 \in \mathscr{A}$ is *asymptotically efficient* for testing $\theta = \theta_0$ against a sequence $\theta_n = \theta_0 + \xi_n^{-1/2}(\theta_0)h$ of local alternatives with $h \neq 0$ if

$$\liminf_{n\to\infty} \pi_n(\theta_n; \phi_n^0) \geq \limsup_{n\to\infty} \pi_n(\theta_n; \phi_n) \,, \qquad \text{for all } \phi_n \in \mathscr{A} \,.$$

From (31) we note that for supercritical branching processes

$$\Lambda_n(\theta_n, \theta_0) = h\Delta_n(\theta_0) + \tfrac{1}{2}h^2W_n(\theta_0) \to 0 \,, \qquad \text{under } \mathbf{P}_{\theta_0}^A \,. \tag{32}$$

Using equation (26) from section 9 we also have that

$$(\Delta_n(\theta_0), W_n(\theta_0)) \to (W^{1/2}(\theta_0)Z, W(\theta_0)) \,, \qquad \text{under } \mathbf{P}_{\theta_0}^A \,. \tag{33}$$

The conditions (32) and (33) constitute the requirements for local asymptotically mixed normal (LAMN) experiments (Le Cam, 1986, chapter 11, contains the general theory). A consequence of this (Basawa and Scott, 1983, Appendix 2 and chapter 1.3) is the following result, which we will not prove.

Proposition 2.13: Assuming the LAMN conditions (32) and (33) for any $h \in \mathbf{R}$, we have that

$$(\Lambda_n(\theta_n, \theta_0), \Delta_n(\theta_0), W_n(\theta_0))$$
$$\xrightarrow{d} \begin{cases} (hW^{1/2}(\theta_0)Z - \frac{1}{2}h^2W(\theta_0), W^{1/2}(\theta_0 Z, W(\theta_0)) & \text{under } \mathbf{P}^A_{\theta_0}, \\ (hW^{1/2}(\theta_0)Z + \frac{1}{2}h^2W(\theta_0), W^{1/2}(\theta_0)Z + hW(\theta_0), W(\theta_0)) & \text{under } \mathbf{P}^A_{\theta_n}. \end{cases}$$

By the Neyman–Pearson lemma we have for any size α_n-test ϕ_n that

$$\pi(\theta_n, \phi_n) \leq \mathbf{P}_{\theta_n}(\Lambda_n(\theta_n, \theta_0) \geq c_n),$$

where c_n is chosen to give the Neyman–Pearson test size α_n. Thus

$$\limsup_{n\to\infty} \pi(\theta_n, \phi_n) \leq \mathbf{P}_{\theta_n}(hW^{1/2}Z + \tfrac{1}{2}h^2W > c_\alpha)$$

where c_α is determined by

$$\mathbf{P}_{\theta_0}(hW^{1/2}Z - \tfrac{1}{2}h^2W > c_\alpha) = \alpha .$$

Since $\Lambda_n(\theta_n, \theta_0)$ is asymptotically equivalent to $h\Delta_n - \frac{1}{2}h^2W_n$ under $\mathbf{P}_{\theta_0}$ we have shown the following.

Theorem 2.9: The test statistic $h\Delta_n - \frac{1}{2}h^2W_n$ is asymptotically efficient.

Remark: A test that is independent of h could reasonably be called *asymptotically uniformly most powerful*. The test in Theorem 2.9 does not have this property in general. However, in the ergodic case the test statistic is asymptotically equivalent to the score statistic Δ_n, which therefore is asymptotically uniformly most powerful against one-sided local alternatives. Swensen (1983) shows that an asymptotically uniformly most powerful test exists if and only if $W(\theta) \equiv 1$. □

To compute the power of the likelihood ratio test, note that

$$J_n^{-1/2}(\theta_0)(\hat{\theta}_n - \theta_0) = W_n^{-1}\Delta_n + o_{\mathbf{P}^A}(1),$$

so the likelihood ratio statistic, from (30), is

$$\lambda_n = 2W_n^{-1}\Delta_n^2 - W_n^{-1}\Delta_n^2 + o_{\mathbf{P}^A}(1) = W_n^{-1}\Delta_n^2 + o_{\mathbf{P}^A}(1)$$

Using Proposition 2.13 we see that

$$\lambda_n \xrightarrow{d} \begin{cases} Z^2, & \text{under } \mathbf{P}^A_{\theta_0}, \\ (Z + W^{1/2}h)^2, & \text{under } \mathbf{P}^A_{\theta_n}. \end{cases}$$

Hence the likelihood ratio statistic is asymptotically χ^2-distributed under the null hypothesis, and a mixture of noncentral χ^2-distributions under local alternatives.

The likelihood ratio statistic $\Lambda(\hat{\theta}_n, \theta_0)$ is, of course, not asymptotically efficient in the sense given above, since it is dominated by $\Lambda(\theta_n, \theta_0)$ at any particular sequence θ_n of local alternatives.

Example: Positive Geometric Offspring

In comparing the score and likelihood ratio tests, Sweeting (1978) and Feigin (1978) looked at the positive geometric offspring distribution. Here we have

$$\lambda_n = \frac{Y_{n-1}}{m(m+1)} (\hat{m} - m)^2 + o_{\mathbf{P}^A}(1) ,$$

$$\Delta_n^2 = W_n \frac{Y_{n-1}}{m(m+1)} (\hat{m} - m)^2$$

and

$$\omega_n = \frac{\hat{m}^n - 1}{\hat{m}(\hat{m} - 1)(\hat{m} + 1)} (\hat{m} - m)^2 .$$

The asymptotic null distributions of the test statistics are those of Z^2, WZ^2, and $W^{-1}Z^2$, respectively, where W is a unit exponential and Z an independent standard normal random variable. Hence the critical values are $a = \chi^2_{1-\alpha}(1)$, $b = \frac{1}{2}\log^2(2\alpha)$, and $c = (1-2\alpha)^2/(2\alpha(1-\alpha))$, respectively. To compute the powers at local alternatives we use Proposition 2.13 and find, as in Sweeting (1978), that

$$\pi(h, \lambda_n) \to \alpha + h(h^2+2)^{-1/2}\Phi(a^{1/2}h(h^2+2)^{-1/2}) \exp(-a(h^2+2)^{-1}) ,$$

$$\pi(h, \Delta_n^2) \to \tfrac{1}{2}(1 + h(h^2+2)^{-1/2}) \exp(-b^{1/2}((h^2+2)^{1/2} - h)) ,$$

and

$$\pi(h, \omega_n) \to \tfrac{1}{2}(1 + (h - c^{1/2})((h - c^{1/2})^2 + 2)^{-1/2}) .$$

The envelope power function (i.e., the power of the Neyman–Pearson test for each particular alternative, which is the asymptotically efficient test from Theorem 2.9) is given by

$$\pi(h, \Lambda(\theta_n, \theta_0)) \to \tfrac{1}{2}\left(1 + \frac{h}{(h^2+2)^{1/2}}\right)\left(\frac{2\alpha}{1 - \dfrac{h}{(h^2+2)^{1/2}}}\right)^{((h^2+2)^{1/2}-h)/((h^2+2)^{1/2}+h)} .$$

Figure 2.4 shows the four power curves for $\alpha = 0.05$. We see that Δ^2 is more powerful than the other two statistics near zero, while λ is better for

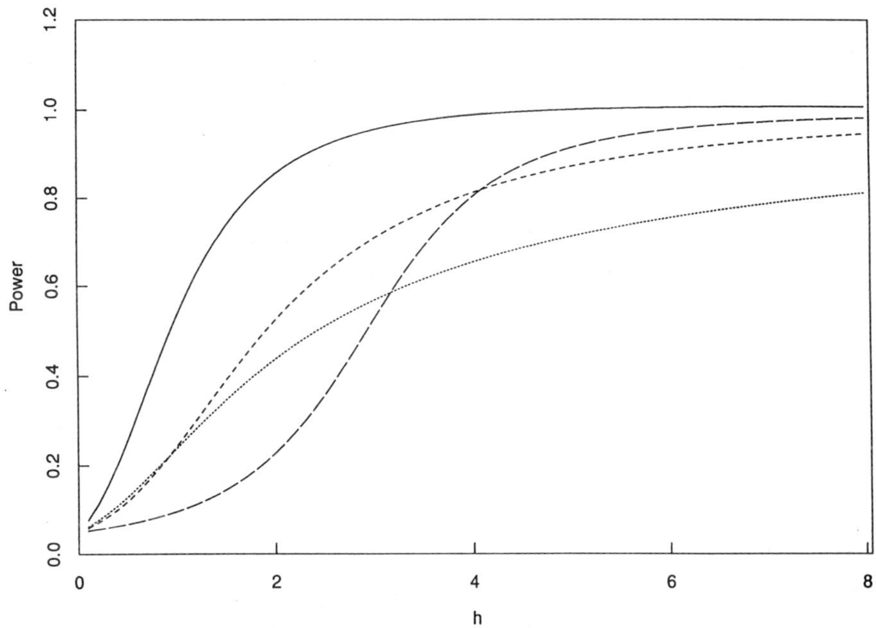

FIGURE 2.4. Power curves for the likelihood ratio test λ_n (short dashes), the score test Δ_n^2 (dots), and the Wald test ω_n (long dashes) in the positive geometric case, testing H_0: $\theta = \theta_0$ as a function of h where the alternative is $\theta_n = \theta_0 + h\xi_n^{1/2}(\theta_0)$. Also shown (solid) is the envelope power curve (the power of the Neyman–Pearson test of θ_0 against θ_n). All tests have asymptotic size 0.05.

moderate h (between 0.91 and 4.1) and ω is best for large h. Since the powers are well below one, at least for $h \leq 6$, these alternatives may all be considered statistically interesting. Also, the power envelope is well above the achieved powers particularly when $h \leq 4$. □

As we discussed in sections 8 and 9, there are reasons to consider conditional inference in this situation. The approach of Fisher–Efron–Feigin suggest inference conditional upon Y_{n-1}, while that of Lauritzen (also favored by Basawa and Scott, 1983) prescribes conditioning upon the "nuisance random variable" W. As we pointed out earlier, the latter approach reduces the model to an ergodic one, since the tail σ-field is generated by W.

Consider the conditional exponential family setup, so that

$$l_n(\lambda) = Y_n \log \lambda - Y_{n-1} \log A(\lambda) - \log \lambda \,. \tag{34}$$

Then (Y_n, Y_{n-1}) is sufficient for the curved exponential family (34) (cf. section 7). In the full exponential model

$$l_n(\boldsymbol{\eta}) = Y_n \eta_1 + Y_{n-1}\eta_2 - K(\boldsymbol{\eta}) \,,$$

it is well known (e.g., Lehmann, 1959, Lemma 8 in section 2.7) that the conditional density of the data, given Y_{n-1}, is free of η_2, so

$$l_n(\boldsymbol{\eta}) = l_n^c(\eta_1 | Y_{n-1}) + \log f_{Y_{n-1}}(Y_{n-1} | \boldsymbol{\eta}) .$$

Conditionality (in the sense of Fisher–Efron–Feigin) requires the use of l_n^c for inference about η_1. As shown by Lehmann (1959, p. 134) the uniformly most powerful unbiased test $\phi_n(Y_n | Y_{n-1})$ of $\eta_1 = \eta_1^0$) against $\eta_1 > \eta_1^0$ (in the full, unconditional model) rejects for large values of Y_n, with the critical value c determined by

$$\mathbf{P}_{\eta_1^0}(Y_n > c | Y_{n-1}) = \alpha .$$

The optimality property of this test cannot be claimed in our situation, since there is no reason to claim that $\eta_2 = \log A(\lambda)$ is a nuisance parameter compared to $\eta_1 = \log \lambda$.

Example: Positive Geometric Distribution

In the geometric case we get

$$f(y_n | y_{n-1}; \log(1-p)) = p(1-p)^{y_n - y_{n-1}} = p(1-p)^{Z_n} ,$$

i.e., a geometric distribution again. We therefore test H_0: $p = p_0$ by rejecting whenever $z_n > c$, where c is chosen as large as possible so that

$$\sum_{k>c} p_0(1-p_0)^k \leq \alpha . \qquad \square$$

Feigin (1986), working from a decision-theoretic point of view, found that W_n is the *deepest correct conditioning* in the sense of Hájek (1967). There is an optimality result, due to Le Cam, which concerns test that are asymptotically conditional size α, defined by

$$\mathbf{E}_{\theta_0}(\phi_n \gamma(W_n)) \rightarrow \alpha \mathbf{E}\gamma(W) \tag{35}$$

for all bounded, continuous functions γ. Again, we will not prove this result (see Feigin, 1986 and Le Cam, 1986, section 11.9).

Theorem 2.10: Among all tests ϕ_n satisfying (35) we have

$$\limsup_{n\rightarrow\infty} \mathbf{E}_{\theta_n} \phi_n \leq \mathbf{P}(Z + hW^{-1/2} \geq z_{1-\alpha} W^{-1/2}) ,$$

so the test rejecting when $J_n^{1/2}(\hat{m}_n - m) \geq z_{1-\alpha}$ is asymptotically optimal in the class (35).

Remark: In the sense of Theorem 2.10 the likelihood ratio test is optimal. $\square$

If we instead want to follow the approach of Lauritzen to find an extreme point model, we must condition on W. In the special case of the positive geometric offspring distribution we discussed this briefly in section 8.

Example: Positive Geometric Distribution

Given $W = w$ the random variables $T_j = Z_j - Z_{j-1}$ are independent Poisson variates with mean $w(m^j - m^{j-1})$. Let $\eta_1 = -\log\theta$ and $\eta_2 = \log(w(1-\theta)/\theta)$. Then the conditional likelihood given $W = w$ is

$$\begin{aligned}\bar{L}_n(\theta) &= \exp\Big(\eta_1 \sum jT_j + \eta_2 \sum T_j - K(\boldsymbol{\eta})\Big) \\ &= \exp(-\eta_1(nZ_n - Y_{n-1}) + \eta_2(Z_n - 1) - K(\boldsymbol{\eta}))\,.\end{aligned}$$

This is another curved exponential family, and the UMPU test of $\eta_1 = \eta_1^0$ against $\eta_1 < \eta_1^0$ rejects for large values of $Y_{n-1} - nZ_n$, the critical constant computed according to the conditional distribution of $\Sigma\, jT_j$ given $\Sigma\, T_j$. Given that $\Sigma\, T_j = t$, the T_j are conditionally distributed as $\mathrm{Mult}_n(t, \mathbf{p})$, where

$$p_j = \frac{(e^{\eta_1} - 1)e^{(j-1)\eta_1}}{e^{n\eta_1} - 1}\,.$$

In this case one can argue that η_2 is more of a nuisance parameter, since it depends on the nuisance random variable W. □

11. RANDOMLY SAMPLED FAMILY SIZES

We now return to the situation described in section 1.6. Assume that we have taken a random sample of r_n individuals in the nth generation, and recorded their family sizes $\mathbf{Y}_n$. If we continue sampling for t_n additional consecutive genrations we get the observation $\mathbf{Y}_{n,t_n}^T = (\mathbf{Y}_n^T, \ldots, \mathbf{Y}_{n+t_n}^T)$. Suppose ξ has the distribution (ip_i/m). Then $\mathbf{E}1/\xi = 1/m$, so a method of moments estimator of m is the harmonic mean

$$\tilde{m} = \left(\frac{1}{R_n}\sum_{j=1}^{R_n}\frac{1}{Y_{n,t_n}^{(j)}}\right)^{-1}$$

where $R_n = \Sigma_{j=0}^{t_n} r_{n+j}$ is the total sample size, and $Y_{n,t_n}^{(j)}$ is the jth entry in $\mathbf{Y}_{n,t_n}$. A straightforward computation, using that asymptotically the sampled individuals are a.s. not siblings, shows that

$$(m^2\sigma_{1/\xi}R_n^{-1/2})(\tilde{m} - m) \xrightarrow{d} \mathbf{N}(0,1)\,,$$

where $\sigma_{1/\xi}^2 = \mathbf{Var}\, 1/\xi$.

If the offspring distribution is exponential family, so $p_i = a_i\theta^i/A(\theta)$, we find that the distribution of ξ is also in the power series class, namely $\mathbf{P}(\xi = i) = ia_i\theta^{i-1}/A'(\theta)$. We can compute an approximate likelihood by assuming that the components of $\mathbf{Y}_{n,t_n}$ are iid with the distribution of ξ. The approximate likelihood is proportional to

$$A'(\theta)^{-R_n}\theta^{S_n - R_n},$$

where S_n is the sum of the observed family sizes. The approximate maximum likelihood estimator is therefore

$$\hat{\theta} = h\left(\frac{S_n}{R_n} - 1\right),$$

where h is the inverse function of $\theta A''(\theta)/A'(\theta)$. Since $m = \theta A'(\theta)/A(\theta)$, the approximate mle is given by the corresponding function of $\hat{\theta}$. Then we get

$$([m'(\theta)h'(\mathbf{E}\xi)]\sigma_\xi/R_n^{1/2})(\hat{m} - m) \xrightarrow{d} \mathbf{N}(0, 1),$$

where $\sigma_\xi^2 = \mathbf{Var}\ \xi$. The approximate mle is asymptotically more efficient than the method of moments estimator, which has asymptotic efficiency

$$e(\theta) = \frac{[m'(\theta)h'(\mathbf{E}\xi)]^2\sigma_\xi^2}{m(\theta)^4\sigma_{1/\xi}^2}.$$

Since $\hat{m}$ is a mle for an iid exponential family, and $\tilde{m}$ is a different estimator, standard theory tells us that $e(\theta) \leq 1$.

Example: Computations of Efficiency

If the offspring distribution is the shifted Bernoulli, with $p_i = \theta^i/(\theta(1 + \theta))$ for $i = 1, 2$, the efficiency is 1. On the other hand, for the positive geometric distribution the efficiency is

$$\left(2\left[\sum_{i=0}^{\infty} \theta^i/(i + 2)\right]^{-1}\right)$$

which varies continuously from 0 ($\theta = 1$) to 1 ($\theta = 0$). □

Application: Estimating Family Size Distributions

The sampling scheme of Burks (1933), where a sampled individual's birth order (rather than the family size) is recorded, is amenable to an analysis similar to that given above. The distribution of observed birth orders is

$$\begin{aligned}\mathbf{P}\{\text{birth order} = j\} &= \sum_{k \geq j} \mathbf{P}\{\text{birth order} = j \mid \text{eventual family size} = k\}\,\frac{kp_k}{m} \\ &= \sum_{k \geq j} \frac{p_k}{m}.\end{aligned}$$

We are observing children, so we will assume that $p_0 = 0$. Then m is the mean number of children *among families with children*. Let N_i be the number of individuals in the sample with birth order i. Note that $\mathbf{E}N_1 = n/m$, where n is the sample size, so we can estimate m using a method of moments approach by

$$\tilde{m} = \frac{n}{N_1} .$$

Since $\mathbf{Var}\, N_1 = n(m-1)/m^2$, a Taylor series expansion yields an approximate standard error for $\tilde{m}$ of $m((m-1)/n)^{1/2}$. We can also estimate the offspring distribution. Noting that $\mathbf{E}(N_k - N_{k+1}) = np_k/m$ we use the estimator

$$\tilde{p}_k = \frac{\tilde{m}}{n}(N_k - N_{k+1})$$

which has an approximate standard error of (Exercise 16)

$$\frac{m}{n}(P_k + P_{k-1} + p_k^2 - 2p_k 1(k=1))^{1/2} \tag{36}$$

where $P_k = \Sigma_k^\infty p_j$. This formula is different from that given in Burks' paper, which does not take the variability of $\tilde{m}$ into account. To assess these estimators, we will look at some data from Burks' paper on birth order of 1800 University of California students enrolled in elementary psychology classes between 1924 and 1929. The data were collected by George M. Stratton, and also contain the present size of each individual's sibling group. Since these students were in their second year of college or above, most sibling groups would be expected to be complete.

Size	N_k	Sibling groups
1	797	275
2	455	457
3	265	440
4	125	247
5	68	147
6	37	92
7	26	56
8	8	30
9	1	20
10	9	13
11	5	10
12	3	8
13	1	4
14	0	1

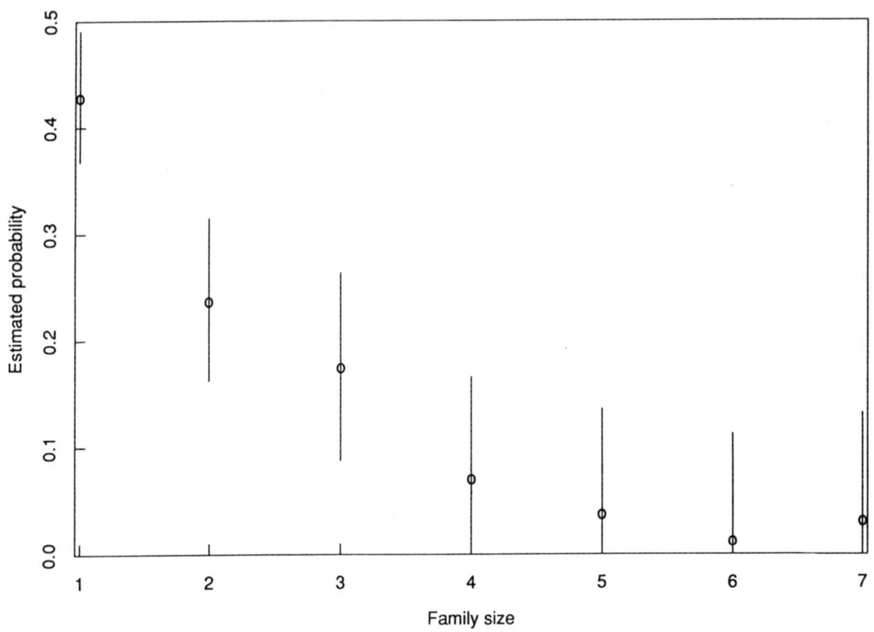

FIGURE 2.5. Estimated family size probabilities for the University of California students. The bars are asymptotic 95% confidence intervals.

We estimate the mean by $\tilde{m} = 1800/797 = 2.26$ with an estimated standard error of 0.06. The mean observed sibling group size is 3.3, but as pointed out earlier this is a biased estimate of mean family size, since small sibling groups have a smaller chance of being included in the sample. We can correct for this by weighting each number proportionally to its sibling group size. The weighted mean, $(\Sigma\, \eta_k)/(\Sigma\, \eta_k/k)$, where η_k is the number of sibling groups of size k, comes out to be 2.32, within one standard error of $\tilde{m}$. This is a limited check on the goodness of fit of the model. Applying $\tilde{m}$ to data on US native white women yields a mean family size of 3.1 children (Keyfitz, 1985, section 11.7), so the California students seem to come from smaller than average families (cf. also Exercise 17). Figure 2.5 shows the estimated family size probabilities. Since the method of moments easily can yield negative estimates when the counts are small, we have grouped the family sizes of 7 or higher. □

12. INFERENCE CONDITIONAL UPON EXTINCTION

Suppose now that we observe extinction. In accordance with the earlier discussion we should make our inferences conditional upon $\{Z_n \to 0\}$ or, equivalently, $\{W = 0\}$, the appropriate tail event. We will assume that the offspring distribution is exponential family. Waugh (1958) derives the conditional probability generating function

$$\mathbf{E}(s^X|W=0) = P(qs;\lambda)/q = P(s;q\lambda).$$

For $m \le 1$, $q = 1$ so we get just the unconditional pgf, but if $m > 1$ we are changing parameter from λ to λq. Note that this brings us to the subcritical domain, since

$$\mathbf{E}(X|W=0) = qP'(qs;\lambda)/q\,|_{s=1} = P'(q;\lambda) < 1.$$

More generally, Daly (1979) looks at probabilities of the form (reminiscent of the partial exponential family in section 3)

$$p_k = a^k g(k, \mathbf{b}) C(a, \boldsymbol{b}),$$

where a and $\mathbf{b}$ are parameters. The offspring probabilities in the conditional model, given extinction, for this model are

$$p_k^* = (aq)^k g(k, \mathbf{b}) C(aq, \mathbf{b}).$$

Again, the parameter a is replaced by aq.

The exponential family likelihood from observing a total of y individuals is proportional to $\lambda^{y-z_0}A(\lambda)^{-y}$. If N is the time of extinction, we see that $Y_N = Y_{N-1} = y$, so in this situation we have a one-dimensional sufficient statistic, which has a distribution belonging to a regular one-parameter exponential family (Becker, 1974). Let $\Lambda_{\le 1} = \{\lambda: 2m(\lambda) \le 1\}$ and $\Lambda_{>1} = \{\lambda: m(\lambda) > 1\}$ be the sub- and supercritical paths of the parameter space, where again $m(\lambda) = \lambda A'(\lambda)/A(\lambda)$. The mle solves $m(\hat{\lambda}) = 1 - z_0/y$, which is always smaller than one. If we assume that $\lambda \in \Lambda_{>1}$, though, the conditional mle solves $m(\hat{q}\hat{\lambda}_{>1}) = 1 - z_0/y$ where $\hat{q}$ solves $P(q; \hat{\lambda}_{>1}) = q$. Thus, the conditional inference we make will depend on what we decide about the parameter value. Since the subcritical conditional mle also maximizes the overall likelihood, we cannot just compare relative likelihoods. Becker (1974) suggests looking at the estimated probability of extinction, namely $\hat{q}^{z_0}$. If this probability is too small, one would reject the possibility of supercriticality. A different approach is to compute an approximate standard error for $\hat{q}$, and appeal to a central limit theorem for the (conditional) mle as z_0, the initial sample size, grows. If $\hat{q}$ is small enough relative to 1, we may conclude supercriticality. An argument much like that in Stigler (1971) shows that

$$\sqrt{z_0}\,(\hat{q} - q) \sim \mathrm{As}\,\mathbf{N}\left(0, \sigma_\lambda^2 \left(\frac{(\partial/\partial\lambda)P(q;\lambda)}{(\partial/\partial q)P(q;\lambda) - 1}\right)^2\right),$$

where σ_λ^2/z_0 is the asymptotic variance of $\hat{\lambda}$ (cf. section 3.5).

Application: Smallpox Epidemics in Europe

The World Health Organization used to publish data on smallpox importations into Europe. Becker (1974) analyzed some of these data. We will look at data from January 1961 to September 1970.

The distribution of Y_∞ for a branching process with geometric offspring distribution can be deduced from queueing theory. Recall from the Example at the end of section 2.6 that the number of customers in an M/G/1-queue arriving during the service time of each customer forms successive generations of a BGW process. If the service time is fixed and taken as a unit of time the offspring distribution corresponds to a Poisson distribution, with mean the arrival rate. If the arrival rate is smaller than 1, the process becomes extinct, i.e., the queue becomes empty) with probability one. The total number of customers until the queue becomes empty was derived by Borel (1942), and is called the Borel–Tanner distribution. The probability function is

$$\mathbf{P}(Y=k)=\frac{1}{(k-1)!}\,k^{k-2}\theta^{k-1}e^{-\theta k}\,.$$

In order for this to be a proper distribution, it is necessary and sufficient that $\theta \le 1$. It has finite mean, $(1-\theta)^{-1}$, if and only if $\theta < 1$.

Fitting a Poisson offspring distribution to the smallpox data (i.e., fitting a Borel–Tanner distribution to the total sizes of each epidemic) we get a χ^2-statistic of 0.54 at 1 df. We have $z_0 = 28$, $y = 391$ so $\hat{\lambda} = \hat{m} = 0.928$. If we assume supercriticality, i.e., that the Europeans were simply lucky to avoid a major epidemic, we get $\hat{q}\hat{\lambda}_{>1} = 0.928$, so $\hat{m}_{>1} = 1.075$ and $\hat{q}^{28} = 0.016$. A large sample 95% confidence interval for $\hat{q}$ is (0.66, 1.06) which covers 1, but it is of somewhat dubious quality as z_0 is rather small. □

13. SOURCES

Bharucha-Reid (1956) proposed the use of branching processes to describe early stages of epidemic processes. Section 1 is based on Bartoszyński (1967) and Becker (1977).

The estimators m^*, $\bar{m}$, and $\hat{m}$ were introduced by Heyde (1975), Lotka (1939), and Harris (1948), respectively. The representations of $\hat{m}_n$ and $\bar{m}_n$ as linear combinations of Z_j/Z_{j-1} are due to Becker (1977).

Basawa (1987) presents material on the conditional least squares approach beyond what is presented in section 3.

The mle for the offspring distribution was obtained by Harris (1948). Section 4 is largely based on Keiding and Lauritzen (1978). In particular, Proposition 2.5 and Lemma 2.2 are from that paper. The Example was given by Keiding (1975a). The application to whooping cranes was discussed by Keiding (1976) and by Nedelman et al. (1987).

Lemma 2.3 is standard (e.g., Nanthi, 1983, Lemma 2.1). Proposition 2.6 is due to Yaglom (1947) and is often called Yaglom's theorem in the literature. The work on stationary measures is from Harris (1963, section 1.11) and the genetical application in section 5 follows the presentation in Moran (1962).

Yanev (1975) proved Theorem 2.2. The martingale central limit theorem of Scott (1978) was preceded by a similar result by Bühler (1969). Theorem 2.3 was first given by Dion (1974), although not quite in the generality presented here. The data for the epidemiology application in section 6 are from Rodrigues da Silva et al. (1963), quoted and analyzed by Becker (1977), and the genetics application in the same section is from Fisher (1958, chapter IV). Proposition 2.9 and Therem 2.4 follow Yanev (1975). The final example was given by Kendall (1951).

The discussion of efficiency is largely based on Heyde and Feigin (1975) and the presentation in Hall and Heyde (1980). The material in Basawa and Scott (1983), Jeganathan (1982), and Swensen (1983) are generalizations of this material to arbitrary nonergodic processes. The theory of estimating equations was first presented by Durbin (1960) and Godambe (1960), and there has been various generalizations to stochastic processes, e.g., Godambe (1985). Godambe and Heyde (1987) prove that the optimal estimators for m and σ^2 are jointly optimal provided the offspring distribution has zero skewness.

Lauritzen developed the theory of extreme point models in his 1982 dissertation, now reprinted as Lauritzen (1988). The presentation in section 8 follows his 1984 survey paper.

The work on ancillarity in the statistical literature has been substantial. The material in section 9 follows Feigin and Reiser (1979) and Sweeting (1986). Equation (25) is due to Basawa and Scott (1976), and derived in more generality by Swensen (1980) and Davies (1985).

Much of the discussion in section 10 follows Basawa and Scott (1983) and Basawa (1981). The work of Feigin (1986), building on work by Le Cam (1986), is also an important source.

Section 11 is based on Maki and McDunnough (1989a,b). The application uses results and data from Burks (1933).

Daly (1979) discusses problems similar to those in section 11 for more general processes. The work of Waugh (1958) and Becker (1974) has been used for the presentation in this section.

EXERCISES

1. Represent m_n^* as a certain average of the $\bar{m}_k$, $k = 1, \ldots, n$.

2. Prove Lemma 2.1.

3. In order to incorporate previous generations while using the idea of the estimator m_n^*, one may study

$$\tilde{m}_n = \left(\prod_1^n Z_i\right)^{2/n(n-1)} .$$

(a) Justify this estimator.
(b) Prove that $\mathbf{E}\tilde{m}_n < m$.
(c) Show that $n(\tilde{m}_n - m) \to 2m \log W$ a.s. on the explosion set.

4. For $a, b, c, d > 0$ and X a nondegenerate positive random variable, show that

$$\mathbf{E}\left(\frac{a + bX}{c + dX}\right) > \frac{a + b\mathbf{E}X}{c + d\mathbf{E}X} - \frac{bd\,\mathbf{Var}\,X}{(c + d\mathbf{E}X)^2} .$$

5. The proof of Proposition 2.5 can be adapted to many different statistics.
(a) Verify that the result is true for $\mathbf{E}Y_1^k$ for any $k \geq 1$.
(b) Does it hold for $\mathbf{Var}\,Y_1$?

6. Suppose P is given by the density

$$p_k = A(\theta)^{-1} a_k \theta^k$$

with $\theta > 0$ and $a_k \geq 0$. This is called a *power series* distribution.
(a) Prove that $P(s) = A(\theta s)/A(\theta)$.
(b) Show that the mean is $m = \theta A'(\theta)/A(\theta)$ and that the variance is $\sigma^2 = ((d/dm)\log\theta)^{-1}$.
(c) Compute the mle of m.

7. Prove that $\hat{m}_n = \Sigma_{r=0}^{\infty} r\hat{p}_r$ where the $\hat{p}_r$ maximize the likelihood based on $Z_0, \ldots, Z_n$, subject to the constraints $p_r \geq 0$ and $\Sigma\, p_r = 1$.
Hint: Show that you need only look at a finite number of indices r, and use a Lagrange multiplier for the second constraint.

8. (a) Using the functional equation in Proposition 2.6 and expanding in powers of s, find expressions for the probabilities q_j in the case of binary splitting, where $P(s) = p_0 + (1 - p_0)s^2$.
(b) Using a probabilistic argument, determine the distribution of Z_n given $Z_n > 0$, and hence the (q_j).

9. (a) For $m > 1$, show that $q_n = q - dP'(q)^n + O(P'(q)^{2n})$, where d is a positive constant.

Hint: Taylor expand P around q to deduce a Lipschitz condition. Let $s = P_n(s)$ and repeat the process to get a bound on $P_n(s) - q$.
(b) For $m < 1$, show that $1 - q_n = O(m^n)$.
Hint: Write $P(s) = 1 - m(1 - s) + r(s)(1 - s)$. Replace s by $P_k(s)$, and multiply the expansions from $k = 0$ to $n - 1$.

10. Prove that if T is the extinction time of a BGW process, then

$$\lim \mathbf{P}(Z_n = j | n + k < T < \infty) = b_j(k) ,$$

where $\Sigma_j b_j(k) = 1$ if $m \neq 1$. Use this result to determine the limiting distribution of $\hat{m}_n$ given that $Z_n > 0$.

11. Suppose that $m = 1$ and $\sigma^2 < \infty$. Prove that

$$\lim_{n \to \infty} \mathbf{P}\left(\frac{2Z_n}{n\sigma^2} > u | Z_n > 0\right) = e^{-u} .$$

12. Let $m = 1$, and suppose that $n/z \to K$, $0 < K \leq \infty$. Show that

$$Y_n(z)/z \xrightarrow{\mathbf{P}} \infty .$$

Hint: Expand the logarithm of the Laplace transform, and verify that

$$R(s) = (2\sigma^{-2}(1 - s))^{1/2} + o((1 - s)^{1/2})$$

as $s \to 1$, where R is the pgf of $Y_\infty = \lim Y_n(1)$.

13. Prove that $\gamma(\theta)$ in Theorem 2.5 must be continuous.

14. Prove that a first order R-efficient estimator is consistent, provided that $I_n(\theta) \to \infty$ a.s. This is called *first order ergodicity* by Lauritzen (1984).

15. (a) Prove that if a transition density of a Markov chain satisfies (24), then the score martingale satisfies (19).
(b) Show that in the conditional exponential family setup, $\hat{\theta}_n$ is consistent for θ provided that $\Sigma_{i=1}^{\infty} H(X_{i-1})$ diverges a.s.

16. (a) Find the bias to order $1/n$ of the estimator $\tilde{m}$ of the Application in section 2.11.
(b) Verify the standard error calculation (36) using a bivariate Taylor expansion of $f(x, y) = xy$.

17. Compare the University of California students in the Application of section 10 to the 1928 US population, given in the Appendix.

18. Using the Nigerian smallpox epidemic data (given in the Appendix), estimate the offspring mean using a Poisson offspring distribution. Use this to assess the proportion that needs vaccination in order to prevent another outbreak. Discuss the assumptions you are making.

CHAPTER 3

Estimating Other Functionals

Having studied the problem of estimating the offspring mean, we now turn to other estimation problems for a BGW process. The offspring variance and the offspring lattice size are found to be consistently estimable when observing only generation sizes from an explosive path. Other parameters of interest, not in general consistently estimable even on explosive paths, are the extinction probability, the offspring distribution, and the age of the process.

1. THE OFFSPRING VARIANCE

There are many situations in which it is important to determine the variance of the offspring distribution. For example, in order to produce confidence intervals for the offspring mean, a consistent estimator of the offspring variance is necessary. In the following application the variance-mean ratio of the offspring distribution turns out to be important.

Application: Photon Detection by Electron Multiplier

Most early devices used for detection and counting of elementary particles and photons, such as the Geiger–Müller counter, depended for their operation upon the creation of a large number of ions in the gas of a counting tube. This had substantial impact on the accuracy of the counts, and could only resolve events more than about 10^{-4} s apart (a phenomenon called *dead time* in counter theory). The vacuum electron multiplying tube was patented by Slepian in 1919 for use in amplifiers. It was important in the development of television as coupling and amplifying unit for cathode-ray tubes (Zworykin et al., 1936). Bay (1941) suggested the use of electron multipliers in electron and photon counting. For this application, the device has a dead time of less than 10^{-8} s, and produces more accurate counts than gas-filled counters. Photon multipliers were used in the detection of the neutrino shower from the recently observed supernova SN 1987a (Genet et al., 1987, chapter 14). An array of photon multipliers was located on top of

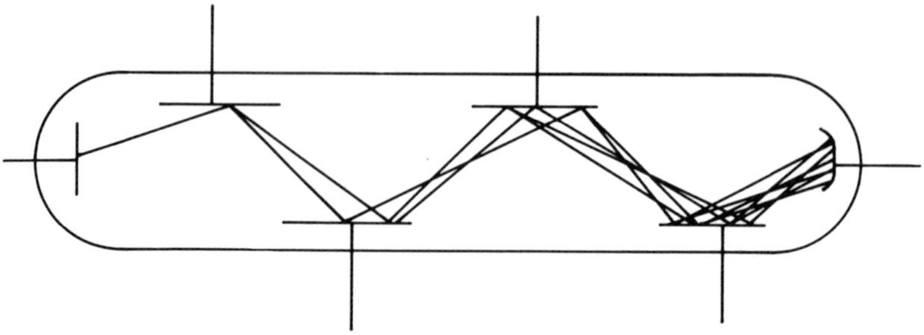

FIGURE 3.1. A schematic depiction of an electron multiplier.

a tank of water in a Japanese mine in order to detect the faint circle of blue light obtained when a neutrino interacts with the water molecules.

The idea behind the electron multiplier is that a primary electron (which in the case of photon detection may be obtained from a photoelectric plate), after being accelerated in an electric field, strikes a flat plate (dynode). The impact produces secondary electrons which in turn are directed towards a second dynode, etc. This direction is usually achieved using a magnetic field. The practical solution to this focusing problem delayed the practical use of the multiplier tube for nearly twenty years. The number of electrons emitted from a dynode is proportional to the bombarding current, the factor of proportionality ranging from a mere fraction to about ten. The value of the ratio of secondary to primary electrons depends on the surface used, and the velocity of the primary electron. For use in electron counting, surfaces with a multiplication factor 4.5–6 at each stage are used, yielding an amplification factor for a ten stage multiplier of 10^6–10^8. Figure 3.1 shows a schematic rendering of an electron multiplier for electron counting use.

The precision of a multiplier is largely dependent on two factors: the noise in the incoming signal, and the noise generated when a primary electron gives rise to secondary elements at each dynode. Suppose that the initial current of primary electrons is i_0, generally a random variable. The resulting output current i_n after n multiplier stages has mean

$$\mathbf{E}i_n = \mathbf{E}i_0\mathbf{E}Z_n = m^n\mathbf{E}i_0 \,, \tag{1}$$

where m is the multiplication factor. If the input current can be assumed independent of the multiplier, which is reasonable over a range of moderate currents, we get

$$\mathbf{Var}\ i_n = \mathbf{E}i_0\ \mathbf{Var}\ Z_n + m^{2n}\ \mathbf{Var}\ i_0 \,. \tag{2}$$

The first term is the noise produced in the secondary emitters, whereas the second is the amplification of the input noise. Zworykin et al. (1936) assumed a Poisson noise in the secondary emission based on the law of small numbers. They proceeded to measure the variance-to-mean ratio and compare this to the calculated noise under the Poisson assumption. From (2) we see that this ratio should be

$$\frac{\mathbf{Var}\ Z_n}{\mathbf{E}Z_n} + m^n \frac{\mathbf{Var}\ i_0}{\mathbf{E}i_0} .$$

If furthermore the input current follows a Poisson process, as would be the case for rectangular pulse shot noise, which is a common model for photoelectric emission, we can compute this ratio theoretically, yielding for a sequence of experiments reported by Zworykin et al. (1936) the following table of results:

No. of stages	m	Measured noise ratio	Derived noise ratio	Offspring variance/ mean ratio
3	3.91	77	80	0.85
3	3.04	40	41	0.92
3	1.90	12.1	12.3	0.96
2	5.45	36.2	36.0	1.03
1	6.00	7.2	7.0	1.20

The close agreement between the third and fourth columns above were used by Zworykin et al. to support the assumptions made. Shockley and Pierce (1938), however, pointed out that the main factor in (2) is the amplified input noise, and therefore the agreement is somewhat exaggerated. The Poisson offspring distribution is, in fact, not a necessary assumption to get this agreement. The last column in the table shows the estimated variance-to-mean ratio for the offspring distribution which differs from unity (the Poisson value) by up to 20%. Other workers (e.g., Kurrelmeyer and Hayner, 1937) found that the Poisson distribution performed well for low voltages, but that the variance at high voltages was considerably larger than what the Poisson theory would predict. □

Perhaps the simplest approach to estimating the variance is to use the residual variance from the least squares criterion:

$$\tilde{\sigma}^2 = \frac{S_n(\hat{m})}{n} = \frac{1}{n} \sum_{k=1}^{n} Z_{k-1}(\bar{m}_k - \hat{m}_n)^2 .$$

This estimator was proposed by Dion (1975). Heyde (1974) proposed a similar estimator, using $\bar{m}_n$ instead of $\hat{m}_n$. In order to deduce the properties of $\tilde{\sigma}^2$, we first look at the case of known m.

Proposition 3.1: Assume that $\mu_4 = \mathbf{E}Z_1^4 < \infty$. Then

$$\left| \frac{1}{n} S_n(m) - \frac{\sigma^2}{n} \sum_1^n 1(Z_{k-1} > 0) \right| \xrightarrow{\text{a.s.}} 0 .$$

Proof: First note that

$$\mathbf{E}D_i \equiv \mathbf{E}(Z_{k-1}(\bar{m}_k - m)^2 - \sigma^2 1(Z_{k-1} > 0) | \mathcal{F}_{k-1}) = 0 ,$$

so that

$$M_n = \sum_1^n D_i = S_n(m) - \sigma^2 \sum_1^n 1(Z_{k-1} > 0)$$

is a zero mean martingale. The variance of a martingale difference term is

$$[2\sigma^4 + (\mu_4 - 3\sigma^4)\mathbf{E}(Z_{k-1}^{-1} | Z_{k-1} > 0)](1 - q_{k-1}) .$$

here we used Lemma 2.1, and the fact that if $\bar{X}$ is the average of n iid, mean μ, random variables, then $\mathbf{E}(\bar{X} - \mu)^4 = 3\sigma^4/n^2 + (\mu_4 - 3\sigma^4)/n^3$. Since $q_k \geq 0$ and $\mathbf{E}(Z_{n-1}^{-1} | Z_{n-1} > 0) \leq 1$, we can bound **Var** M_n by $n(\mu_4 - \sigma^4)$, from which weak consistency follows. The strong convergence follows from the martingale law of large numbers (Lemma 2.5). We use the trivial choice $\mu_n = n$, which yields the usual form of the law of large numbers, and must verify that

$$\sum i^{-2}\mathbf{E}(D_i^2 | \mathcal{F}_{i-1}) < \infty ,$$

which is clear from the estimates above. □

Corollary

$$S_n(m)/n \xrightarrow{\text{a.s.}} \begin{cases} \sigma^2 , & \text{on } A, \\ 0 , & \text{elsewhere} . \end{cases}$$

Proof: Immediate from Proposition 3.1. □

The next result deals with the asymptotic normality of $S_n(m)$.

Proposition 3.2: Assume that $\mu_4 < \infty$. Then

$$\left(\frac{n}{2\sigma^4}\right)^{1/2} \left(\frac{1}{n} \sum_{k=1}^n Z_{k-1}(\bar{m}_k - m)^2 - \sigma^2\right) \xrightarrow{d} \mathbf{N}(0, 1) \qquad \text{on } A .$$

Proof: Let $D_j = (Z_j - mZ_{j-1})^2/Z_{j-1} - \sigma^2 1(A)$. We need the following martingale central limit theorem, given in Hall and Heyde (1980, Corollary 3.1).

Lemma 3.1: Let $(\zeta_{ni}, \mathcal{F}_{ni}, 1 \leq i \leq k_n, n \geq 1)$ be a zero-mean square integrable martingale array with differences ξ_{ni}, and let η^2 be an a.s. finite random variable. Suppose that

$$V_{nk_n}^2 = \sum \mathbf{E}(\xi_{ni}^2 | \mathcal{F}_{n,i-1}) \xrightarrow{\mathbf{P}} \eta^2$$

and that for all $\epsilon > 0$

$$\sum_i \mathbf{E}(\xi_{ni}^2 1(|\xi_{ni}| > \epsilon | \mathcal{F}_{n,i-1}) \xrightarrow{\mathbf{P}} 0 .$$

Then if $\mathcal{F}_{ni} \subseteq \mathcal{F}_{n+1,i}$ for $1 \leq i \leq k_n$, $n \geq 1$, the random variables ζ_{nk_n} converge stably to a random variable Z with characteristic function $\mathbf{E} \exp(-\frac{1}{2}\eta^2 t^2)$.

We will use $\mathcal{F}_{ni} = \mathcal{F}_i^Z$, so that the σ-field inclusion part is trivial, and $\xi_{ni} = s_n^{-1} D_i$, whereas $s_n^2 = \Sigma_1^n \mathbf{E} D_j^2$. A moment computation in the proof of Proposition 3.1 shows that

$$\frac{1}{n} \sum_1^n \mathbf{E}(D_j^2 | \mathcal{F}_{j-1}) \xrightarrow{\text{a.s.}} 2\sigma^4 , \qquad \text{on } A .$$

It is easy to verify that

$$\frac{s_n^2}{n} \xrightarrow{\text{a.s.}} 2\sigma^4 , \qquad \text{on } A ,$$

from which it follows that

$$s_n^{-2} \sum_1^n \mathbf{E}(D_j^2 | \mathcal{F}_{j-1}) \xrightarrow{\text{a.s.}} 1(A) .$$

Next we need to verify the Lindeberg condition

$$s_n^{-2} \sum_{j=1}^n \mathbf{E}(D_j^2 1(|D_j| > \epsilon s_n) | \mathcal{F}_{j-1}) \xrightarrow{\text{a.s.}} 0 .$$

It suffices to show that $\mathbf{E} D_n^2 1(|D_n| > c\sqrt{n}) \to 0$ for all $c > 0$. For large n

$$\begin{aligned} \mathbf{E} D_n^2 1(|D_n| > c\sqrt{n}) \leq 2\mathbf{E} Z_{n-1}^{-2} V_{Z_{n-1}}^4 1(Z_{n-1}^{-2} V_{Z_{n-1}}^4 > c^2 n) \\ + 2\sigma^4 \mathbf{P}\{V_{Z_{n-1}}^2 > c\sqrt{n} Z_{n-1}\} , \end{aligned}$$

where $V_n = \Sigma_1^n X_i$ and the X_i are iid with the centered offspring distribution. The convergence of the first term follows if we can show uniform integrability, and the second term converges to zero by Markov's inequality. The uniform integrability is a consequence of the following result (Billingsley, 1968, Theorem 5.4).

Lemma 3.2: Let X_i be positive, integrable random variables such that $X_i \xrightarrow{d} X$. Then if $\mathbf{E}X_i \to \mathbf{E}X$, the X_i are uniformly integrable.

To verify the condition of the Lemma, notice that $\mathbf{E}V_n^4/n^2 = 3\sigma^4 + \mu_4/n \to 3\sigma^4$, and that $V_n/\sqrt{n} \to \mathbf{N}(0, \sigma^2)$. □

We now proceed to study the estimator $\tilde{\sigma}^2$. The consistency, under the assumptions of Proposition 3.1, is an easy consequence of the following result.

Proposition 3.3: Let $0 < \sigma^2 < \infty$. Then

$$S_n(m) - S_n(\hat{m}) \begin{cases} \xrightarrow{d} \chi^2(1)\,, & \text{on } A \\ \xrightarrow{\mathbf{P}} Y_\infty\left(1 - \dfrac{1}{Y_\infty} - m\right)^2\,, & \text{on } A^c\,. \end{cases}$$

Proof

$$(S_n(m) - S_n(\hat{m})) = \sum_{k=0}^{n-1} [(m^2 - \hat{m}^2)Z_k + 2(\hat{m} - m)Z_{k+1}]$$

$$= (\hat{m} - m) \sum_{k=0}^{n-1} [2Z_{k+1} - Z_k(\hat{m} + m)] = Y_{n-1}(\hat{m} - m)^2\,.$$

Using Theorem 2.3 we see that $Y_{n-1}(\hat{m} - m)^2 \xrightarrow{d} \chi_1^2$ on A. The case A^c is similar. □

Remark: It follows from the proof above that $\tilde{\sigma}^2$ will tend to underestimate σ^2 on A, since it is strictly smaller than $S_n(m)/n$, which is an unbiased estimator of σ^2. Since the difference $S_n(m) - S_n(\hat{m})$ has mean 1, it may make sense to look at the bias-corrected estimator $S_n(\hat{m})/(n-1)$. As a crude approximation, if we say that $\mathbf{Var}(S_n(\hat{m})/n) \approx 2\sigma^4/n$, the approximate mean squared error decreases with the bias-correction whenever

$$\frac{1}{n^2} + \frac{2\sigma^4}{n} \geq \frac{2\sigma^4}{n} \frac{n^2}{(n-1)^2}\,,$$

i.e., if $\sigma^2 > \frac{1}{2}$, or if $\sigma^2 \leq \frac{1}{2}$ and $n \leq 1/(1 - 4\sigma^4)$. □

We can now summarize the properties of the variance estimator $\tilde{\sigma}^2$.

Theorem 3.1: Suppose that $\mu_4 < \infty$. Then

$$\mathbf{E}\tilde{\sigma}^2 < \sigma^2\,,$$

$$\tilde{\sigma}^2 \xrightarrow{\mathbf{P}} \begin{cases} \sigma^2\,, & \text{on } A\,, \\ 0\,, & \text{on } A^c\,. \end{cases}$$

In addition

$$\left(\frac{n}{2}\right)^{1/2}(\tilde{\sigma}^2 - \sigma^2) \xrightarrow{d} \mathbf{N}(0, \sigma^2)\,, \qquad \text{on } A\,.$$

Proof: The theorem follows from Propositions 3.1, 3.2, and 3.3. □

Remarks: (i) Heyde (1974) proved the result of Theorem 3.1 for the estimator $\bar{\sigma}_n = S_n(\bar{m}_n)/n$. Simple algebra shows that

$$\tilde{\sigma}_n - \bar{\sigma}_n = Y_{n-1}(\hat{m}_n - \bar{m}_n) \times (\hat{m}_n^2 + \bar{m}_n^2 - 2\hat{m}_n)/n\,.$$

The second term on the right-hand side goes to zero in probability by the consistency of $\bar{m}$ and $\hat{m}$, and the first term goes to zero in probability since the two estimators are both first order efficient.

(ii) The maximum likelihood estimator of the offspring variance based on observing the whole family tree is not the same as the estimator above. In fact, Dion (1975) shows that the mle from the whole family tree converges to normality at the rate $Y_{n-1}^{1/2}$, rather than $n^{1/2}$. We will consider the maximum likelihood estimator based on observing only generation sizes in section 3 below. □

One reason to consider the estimator $S_n(m)/n$ is the following: for each k, $\tau_k = Z_k(\bar{m}_k - m)^2$ is an unbiased conditional method of moment estimator. Hence one can combine these estimators to obtain one with more information. Looking at linear estimators, an easy computation shows that $\mathbf{Var}(\tau_k|Z_{k-1}) = 3\sigma^4 + (\mu_4 - 3\sigma^4)/Z_{k-1}$. As Z_{k-1} becomes large, the first term dominates, and we are led to equal weighting.

This estimator also arises from a limiting argument. If we think of the process as a Markov chain with transition probabilities $p(z_k|z_{k-1})$ we can use a local limit theorem on the transition probabilities on $(Z_n \to \infty)$, yielding the density of a $\mathbf{N}(mZ_{k-1}, \sigma^2 Z_{k-1})$. $S_n(m)/n$ is the mle of σ^2 in this approximate model. If we also estimate m by maximum likelihood, we get the estimators $\hat{m}$ and $\tilde{\sigma}^2$. This was the approach by Eschenbach and Winkler (1975) and Duby and Rouault (1982).

2. ESTIMATING THE LATTICE

We have found consistent estimators of m and σ^2 on the explosion set. The remaining possibly estimable functionals are the lattice size d and the lattice offset h. In this section we will discuss the estimation of the lattice size. We do this mainly for completeness, as one is unlikely to encounter this estimation problem in practice. The discussion in this section has some bearing on section 4, where we discuss the properties of the maximum likelihood estimator of the offspring variance.

As pointed out in Exercise 1.10, the lattice offset cannot be determined if the process starts from two ancestors. In order to estimate the lattice size, assume that $Z_0 = 1$. Let

$$S_1 = \{l: i + jl = Z_1, \text{ some } i, j \geq 0\} .$$

These are the possible lattice sizes after observing Z_1. Now define recursively

$$S_k = \{l \in S_{k-1}: Z_{k-1}i + jl = Z_k, \text{ some } i, j \geq 0, i, l \text{ consistent with } Z_1, \ldots, Z_{k-1}\} .$$

Here the pair i, l is consistent with $Z_1, \ldots, Z_{k-1}$ if $Z_j \equiv i^j \bmod l$ for $j = 1, \ldots, k-1$. Suppose that $Z_k \to \infty$. Define $\hat{d}_n = \max S_n$ for $n > 1$, and define $\hat{d}_1 = \infty$, with the corresponding offset $h_1 = Z_1$. h_1 is an upper bound to the true offset. Let k be the first integer such that $Z_k \neq Z_1^k$. Then we must have $d \leq Z_k$, which is also an upper bound for $\hat{d}_k$. Clearly, k will be finite with probability one, provided that the offspring distribution is not concentrated on one point only. Since the $\hat{d}_n$ form a decreasing sequence, $\hat{d}_n \to d_\infty$ with probability one, where $d_\infty \geq d$. Let $h_n = Z_1 \bmod \hat{d}_n$. Clearly, $h_n \to h_\infty$ a.s.

Example: The Computation of S_k

The following path was generated from the distribution $P(s) = 0.7 + 0.05s^2 + 0.25s^4$, starting from $Z_0 = 1$:

4 8 12 16 20 20 28 32 54 44 .

To compute S_1, note that $h_1 = 4$. Looking at possible values of i we get

i	l
0	1, 2, 4
1	1, 3
2	1, 2
3	1
4	≥1

whence $S_1 = \mathbf{Z}_+$. Continuing to S_2, we see that $Z_2 \neq Z_1^2$. Hence $d \leq 8$. More specifically, again separating on possible values of i, we get get

i	l
0	1, 2, 4, (8)
1	1, (2), (4)
2	1, 2, (≥3)

Values of i greater than 2 are impossible, and values of l that are incompatible with Z_1 are in parentheses. Hence $S_2 = \{1, 2, 4\}$. The same is the case for $k = 3, \ldots, 8$, while $S_9 = S_{10} = \{1, 2\}$, from which $d_{10} = 2$ and $h_{10} = 0$. □

We now need a technical result.

Lemma 3.3: Let $V_n = \Sigma_1^n U_i$ where the U_i are iid with lattice size d, lattice offset h, mean μ and variance $\sigma^2 < \infty$. Define M by

$$\frac{V_n - nh}{d} \equiv M \bmod r .$$

Then, as $n \to \infty$, $M \xrightarrow{d} U\{0, \ldots, r-1\}$.

Proof: Without loss of generality let $d = 1$, $h = 0$. By the local limit theorem (Lemma 1.4) we have that

$$\left| \sqrt{n}\mathbf{P}\left\{\sum_1^n U_i = k\right\} - \frac{1}{\sigma}\phi\left(\frac{k - n\mu}{\sigma\sqrt{n}}\right)\right| \to 0 \qquad \text{uniformly in } k .$$

Let $B_n(a) = \{k\colon |(k - n\mu)/(\sigma\sqrt{n}) \le a\}$. Then

$$\begin{aligned}
\mathbf{P}\left\{\sum_1^n U_i \equiv j \bmod r, \sum_1^n U_i \in B_n(a)\right\} &= \sum_{l:\, rl+j \in B_n(a)} \mathbf{P}\left\{\sum_1^n U_i = rl + j\right\} \\
&= \frac{1}{\sigma\sqrt{n}} \sum_{l:\, rl+j\in B_n(a)} \phi\left(\frac{rl + j - n\mu}{\sigma\sqrt{n}}\right) + \epsilon_n \\
&= \frac{1}{r}\left(\Phi(B_n(a)) + \delta_n(a)\right) + \epsilon_n ,
\end{aligned}$$

where $\epsilon_n \to 0$ uniformly in l by the local limit theorem, and $\delta_n(a)$ is the error in approximating $\Phi(B_n(a))$ by its Riemann sum, so for any a, $\delta_n(a) \to 0$. We now choose a large, then n large, to make the probability arbitrarily close to $1/r$. □

Notice that

$$Z_k = hZ_{k-1} + d \sum_1^{Z_{k-1}} \left(\frac{X_{k,i} - h}{d}\right).$$

We are interested in possible lattice sizes $r \ge d$, and will try to prove that if $r > d$, the path will eventually tell us so. In order for the pair (r, h^*), where $h^* = Z_1 \bmod r$, to be consistent with $(Z_1, \ldots, Z_n)$ we must have that $Z_{k-1} \equiv (h^*)^{k-1} \bmod r$. Thus the following equation must hold for $k = 1, \ldots, n$:

$$h(h^*)^{k-1} + dU_k \equiv h^* Z_{k-1} \bmod r , \tag{3}$$

where $U_k = \Sigma_1^{Z_{k-1}} (X_{k,i} - h)/d$. Assuming explosion, U_k is asymptotically uniform by the lemma. The first term on the left-hand side of (3) cycles through a list of mod r residue classes. As to the second term, the values 0 and d are both possible values if $r > d$, but they belong to different residue classes. Thus

$$\mathbf{P}\{(3) \text{ fails for } k = n | (3) \text{ holds for all } k < n\} \geq \min(\mathbf{P}\{U_n = 0\}, \mathbf{P}\{U_n = 1\}) \geq \frac{1}{r} + \epsilon_n ,$$

where $\epsilon_n \to 0$, and we can take the sequence decreasing. More formally, then, define

$$C_n^r = \bigcap_{k=1}^{n} \{r \text{ divides } Z_k - h^* Z_{k-1}\} .$$

Notice that C_n^r is necessary for $Z_1, \ldots, Z_n$ to be consistent with r. Given b with $1 < b < m$ and k_0 large enough, define

$$A_n = C_n^r \cap \{Z_j > b^j, k_0 < j \leq n\} .$$

By the standard conditioning argument for Markov chains

$$\mathbf{P}(A_{n+1}) = \mathbf{E}[\mathbf{P}^{Z_n}\{Z_1 = h^* Z_0 + rU_1\} 1(A_n) 1(Z_{n+1} > b^{n+1})] .$$

By our earlier argument

$$\mathbf{P}^k(dU_1 \equiv (h - h^*)k) \leq 1 - \frac{1}{r} + \epsilon_k ,$$

so, e.g., $\mathbf{P}(A_{n+1}) \leq (1 - 1/2r)\mathbf{P}(A_n)$ for all n large enough, whence $\mathbf{P}(A_n) \to 0$ as $n \to \infty$.

Furthermore, by the argument used in proving Proposition 1.4,

$$\mathbf{P}\{(A_k \,\Delta\, C_k^r) \cap \{\text{explosion}\}\} \to 0 ,$$

where Δ stands for symmetric difference. Hence if $r > d$, the probability that $(Z_1, \ldots, Z_n)$ is consistent with r goes to zero on the explosion set. Thus, eventually, $\hat{d}_n = d$. We have proved the following result.

Theorem 3.2: Conditional upon nonextinction, if $Z_0 = 1$, then $\hat{d}_n$ converges a.s. to the true lattice size. The lattice offset can be read off Z_1.

3. THE OFFSPRING DISTRIBUTION

Even though the results in Chapter 1 indicate that there are no consistent nonparametric estimators of the offspring distribution (not even on the explosion set), it is still a relevant parameter to estimate. For example, a nonparametric maximum likelihood estimator of the offspring distribution yields a maximum likelihood estimator of the variance σ^2 (section 4). The likelihood of the observations was given by Eschenbach and Winkler (1975) as

$$L_n = \prod_{i=0}^{n-1} p(Z_{i+1}|Z_i)\,, \tag{4}$$

where we compute $p(j|i)$ as

$$p(j|i) = \sum \frac{i!}{e_0!e_1!\cdots e_j!}\, p_0^{e_0} p_1^{e_1} \cdots p_j^{e_j}\,, \tag{5}$$

where the sum is over all $(j+1)$-tuples $(e_0, e_1, \ldots, e_j)$ for nonnegative integers e_k such that $\Sigma_{k=0}^{j} e_k = i$ and $\Sigma_{k=0}^{j} ke_k = j$. This is equivalent to summing over all partial family trees for which the first n generation sizes take on the values $Z_1, Z_2, \ldots, Z_n$. Denote the number of such trees by $L = L(Z_1, \ldots, Z_n)$. Also denote by $n_{m,j}^{(l)}$ the number of individuals of the mth generation with exactly j offspring in the lth of these trees. Let $n_j(l) = \Sigma_{m=0}^{n-1} n_{m,j}^{(l)}$. Dion et al. (1982) present a representation theorem for the nonparametric mle of an offspring distribution, supported on $(0, \ldots, K)$ for known K.

Proposition 3.4: The mle for the offspring distribution $p_0, \ldots, p_k$ satisfies

$$\hat{p}_k = \sum_{l=1}^{L} \frac{n_k(l)}{Y_{n-1}}\, \hat{\nu}_l\,,$$

where

$$\hat{\nu}_l = \frac{\prod_{i=0}^{K} \hat{p}_i^{n_i(l)}}{\sum_{k=1}^{L} \prod_{i=0}^{K} \hat{p}_i^{n_i(k)}}\,.$$

Proof: From (4) and (5) above we see that the likelihood can be written

$$L_n = \sum_{l=1}^{L} \prod_{i=0}^{K} p_i^{n_i(l)}\,.$$

We maximize the constrained likelihood

$$H_n = L_n + \lambda\left(1 - \sum_{i=0}^{K} p_i\right).$$

Define $G_n = \{k: n_k(l) > 0 \text{ for some } l\}$. Then for all $k \in G_n$ we have

$$\frac{\partial}{\partial p_k} H_n = \frac{1}{p_k} \left(\sum_{l=1}^{L} n_k(l) \sum_{i=0}^{K} p_i^{n_i(l)} \right) - \lambda \, .$$

Setting the partial derivatives equal to zero we get

$$\hat{p}_k = \frac{1}{\lambda} \left(\sum_{l=1}^{L} n_k(l) \prod_{i=0}^{K} p_i^{n_i(l)} \right) .$$

Notice that if for some k $n_k(l) = 0$ for all l we must have $\hat{p}_k = 0$, since otherwise the likelihood would be zero $(0^0 = 1)$. Also notice that $\Sigma_{k=0}^{K} n_k(l) = Y_{n-1}$. Thus we have

$$\lambda = Y_{n-1} \sum_{l=1}^{L} \prod_{i=0}^{K} p_i^{n_i(l)} ,$$

from which the representation follows. □

Notice that $\hat{p}_k$ is a weighted average of the $n_k(l)/Y_{n-1}$, with the weights being the estimated relative likelihood of the lth tree. As shown in section 2.4, if one can observe the entire family tree, the mle of p_k is just n_k/Y_{n-1}. Proposition 3.4 suggests a recursive algorithm for computing the mles. However, this requires the enumeration of all trees that could have yielded the sample. This may be quite a formidable task. For example, even if the offspring distribution is only supported on $\{0, \ldots, 4\}$, there are 342,282 different possible trees representing the first five generations (Latour, 1981).

Example: An mle Calculation

To illustrate an instance of the maximum likelihood estimate of the offspring distribution, consider a simple case where $Z_1 = 3$, $Z_2 = 2$, and $Z_3 = 1$. Then $Y_2 = 6$, and there are two possible trees, as illustrated in Figure 3.2.

The weights are given by

$$\hat{\nu}_1 = \frac{\hat{p}_1^2}{\hat{p}_1^2 + \hat{p}_0 \hat{p}_2}$$

and

$$\hat{\nu}_2 = \frac{\hat{p}_0 \hat{p}_2}{\hat{p}_1^2 + \hat{p}_0 \hat{p}_2} .$$

FIGURE 3.2. The two possible trees generating the path $(Z_1, Z_2, Z_3) = (3, 2, 1)$.

It is easy to check that the solution is given by $\hat{p}_0 = 2/6$, $\hat{p}_1 = 3/6$, $\hat{p}_2 = 0$, and $\hat{p}_3 = 1/6$. Thus, the method of maximum likelihood selects tree 1 as that which has generated the data, and rules tree 2 out completely. □

An easy consequence of the representation in Proposition 3.4 is the following.

Corollary: The mle of m, based on $Z_0, \ldots, Z_n$ is

$$\hat{m}_n = \sum_{k=1}^{K} k\hat{p}_k = \frac{Y_n - Z_0}{Y_{n-1}} .$$

Proof: Notice that

$$\sum_{k=0}^{K} kn_k(l) = \sum_{m=0}^{n-1} \sum_{k=0}^{K} kn_{m,k}^{(l)} = Z_1 + \cdots + Z_n . \qquad \square$$

This is a third way of proving that $\hat{m}$ is the mle based on observing only generation sizes, as soon as we notice that we can choose K to be the largest observed generation size, using the same argument that allowed us to go from finite to infinite support in the proof of Proposition 2.4.

We now outline a computationally more feasible algorithm for computing the mle of the offspring distribution. This uses Harris' form of the estimator, but instead of weighting the Harris estimators with respect to the relative likelihood of different possible trees, we will compute the conditional expectation of $n_j = \Sigma_{m=1}^{n} \Sigma_{l=1}^{Z_m-1} 1(X_{m,l} = j)$, given $Z_0, \ldots, Z_n$. Let $K_n = \max(Z_i; i \leq n)$. Write

$$\hat{n}_j = \mathbf{E}(n_j | \mathcal{F}_n) .$$

This is a function of $p_0, \ldots, p_{K_n}$. Select a starting value $P^{(0)} = (p_0^{(0)}, \ldots, p_{K_n}^{(0)})$. Now compute $\hat{n}_j(P^{(0)})$ as if $P^{(0)}$ were the true distribution. Use the $\hat{n}_j(P^{(0)})$ to compute a new estimate $P^{(1)}$ by the Harris formula,

$$p_k^{(1)} = \hat{n}_k(P^{(0)}) / Y_{n-1} .$$

That is, replace the true value n_k by your best current guess, $\hat{n}_k$, and use that guess to re-estimate the probabilities as if we knew the true n_k. Iterate this procedure until it converges. The results of Dempster et al. (1977) and Wu (1983) can be applied to demonstrate that this algorithm (which is a special case of the strategy called the EM-algorithm) increases the likelihood at every iteration. We do not currently know much about the modality of the likelihood. The algorithm is somewhat sensitive to starting values, but starting from a Poisson distribution truncated above at K_n with parameter $\hat{m}$ appears to work well in a variety of situations.

Let $n_{k,j} = \Sigma_{i=1}^{Z_{k-1}} 1(X_{k,i} = j)$. Then $\hat{n}_j = \Sigma_{k=1}^{n} \mathbf{E}(n_{k,j} | \mathcal{F}_n)$. The pertinent computation is

$$\mathbf{E}\left(\sum_{i=1}^{Z_{k-1}} 1(X_{k,i} = j) \middle| \mathcal{F}_n \right) = z_{k-1} \mathbf{P}(X_{k,i} = j | \mathcal{F}_n) .$$

By independence we need only compute

$$\mathbf{P}(\xi_1 = j | \xi_1 + \cdots + \xi_l = L) ,$$

where the ξ_i are iid (p_i). Writing $p^{*r}(k) = \mathbf{P}(\xi_1 + \cdots + \xi_r = k)$ we get

$$\mathbf{P}(\xi_1 = j | \xi_1 + \cdots + \xi_l = L) = p_j \frac{p^{*(l-1)}(L-j)}{\sum_{i \le L} p^{*(l-1)}(L-i) p_i} . \tag{6}$$

We use $L = Z_k$ and $l = Z_{k-1}$. The resulting algorithm is

$$\hat{p}_j^{(l+1)} = \frac{\sum_{k=1}^{n} Z_{k-1} \dfrac{(\hat{P}^{(l)})^{*(Z_{k-1}-1)}(Z_k - j)\hat{p}_j^{(l)}}{(\hat{P}^{(l)})^{*Z_{k-1}}(Z_k)}}{Y_{n-1}} .$$

This formula is easily computed using a discrete convolution algorithm, such as the discrete Fourier transform. For large values of Z_k, a normal approximation to the convolution (using the mean and variance from the previous iteration) is accurate. It is worth noting (Exercise 2) that the mean of the estimated distribution at each iteration step is $\hat{m}$. The use of an Aitken acceleration (e.g., Press et al., 1986, formula 5.1.4; Melijson, 1989) improves the speed of the algorithm, but does not always converge to the same local maximum as the original method.

Example: Binary Splitting

In special parametric cases we can find a closed form for the conditional probability in (6). In the case of binary splitting, i.e.,

$$\xi_i = \begin{cases} 0 , & \text{with probability } p , \\ 2 , & \text{with probability } 1 - p , \end{cases}$$

we get

$$\mathbf{P}(\xi_1 = 2 | \xi_1 + \cdots + \xi_l = L) = \frac{\binom{l-1}{L/2-1} p^{L/2}(1-p)^{l-L/2}}{\binom{l}{L/2} p^{L/2}(1-p)^{l-L/2}} = \frac{L}{2} l .$$

Thus, the conditional mean

$$\mathbf{E}(n_{k,2}|\mathcal{F}_n) = Z_{k-1}\frac{Z_k}{2Z_{k-1}} = \frac{Z_k}{2} = n_{k,2}$$

is the true value (which we, of course, can read off the path). □

4. MAXIMUM LIKELIHOOD ESTIMATION OF THE OFFSPRING VARIANCE

Define

$$\hat{\sigma}^2 = \sum_{k=0}^{K} \hat{p}_k(k - \hat{m}_n)^2 .$$

Then, by the invariance of maximum likelihood estimation, $\hat{\sigma}^2$ is a maximum likelihood estimator of the offspring variance.

It is difficult to study directly the consistency properties of the estimator $\hat{\sigma}^2$, since there is no closed form for $\hat{p}_k$, and since these estimators by themselves are not consistent. An indirect approach is to study BGW processes with distinct offspring variances but the same offspring mean and the same lattice. If the likelihood ratio of two such processes, each starting from n individuals, can be shown to diverge to zero, uniformly over all BGW measures with offspring variance separated by a given amount, this suffices to verify the consistency in question.

Theorem 3.3: Let P be a fixed offspring distribution on $\{1, \ldots, N\}$ with $p_i > \delta' > 0$, and let $\mathcal{Q}$ be the class of offspring distributions Q on $\{1, \ldots, N\}$ such that $q_i > \delta'$, and $|\sigma_Q^2 - \sigma_P^2| > \delta > 0$. Then, writing $L_{a,b}(R)$ for the likelihood of $Z_a, \ldots, Z_b$ (the conditional distribution of $Z_{a+1}, \ldots, Z_b$ given $\mathcal{F}_a$) when the offspring distribution is R, we have

$$\inf_{Q\in\mathcal{Q}} \frac{L_{0,n}(P)}{L_{0,n}(Q)} \to \infty \text{ a.s. } (\mathbf{P}) ,$$

where $\mathbf{P}$ is the branching process measure corresponding to the offspring distribution P.

Proof: By the law of the iterated logarithm for random walks and the representation in Exercise 1.6, we have that

$$\left|\frac{Z_{k+1} - mZ_k}{(Z_k \log\log Z_k)^{1/2}}\right| \leq M \text{ a.s. for } k \text{ large enough} \tag{7}$$

and by the proof of Proposition 1.4 we can ensure that

$$Z_k > c\theta^k \tag{8}$$

for some $1 < \theta < m$ with arbitrarily high probability. Hence, with probability greater than $1 - \epsilon$ we can make the inequalities both in (7) and (8) hold for all k. Henceforth we only consider such paths.

Let $k < n$. By the local central limit theorem (Lemma 1.4) we can approximate $L_{k-1,k}(P)$ by $A_{k-1,k}(P) = \phi((Z_k - mZ_{k-1})/\sigma_P Z_{k-1}^{1/2})$, and hence we can approximate $L_{k,n}(P)$ by $A_{k,n}(P) = \Pi_{k+1}^n A_{i-1,i}(P)$. Note that for positive a_i, b_i

$$\prod \left(\frac{a_i}{b_i} - 1\right) \le \left|\exp\left(\sum \left(\frac{a_i}{b_i} - 1\right)\right) - 1\right| .$$

Since the error in the local limit theory is $O(Z_{n-1}^{-1/2})$, we have

$$\left|\frac{L_{k,n}(P)}{A_{k,n}(P)} - 1\right| \le \left|\exp\left(\sum_{j=k+1}^{\infty} \frac{c}{Z_j^{1/2} A_{j-1,j}(P)}\right) - 1\right| < \infty$$

on the set of paths considered. Hence $L_{k,n}(P)/A_{k,n}(P) \to 1$ a.s. (**P**) as $k \to \infty$. A careful study of the proof of Lemma 1.4 reveals that the result is uniform in $\mathcal{Q} \cup \{P\}$, since the lattice sizes are the same for all members of this class. Hence

$$\lim_{k\to\infty} \sup_{n\ge k} \left| \inf_{Q\in\mathcal{Q}} \frac{L_{k,n}(Q)}{A_{k,n}(Q)} \to 1 \right| = 0 \text{ a.s.}$$

Now write

$$\inf_{\mathcal{Q}} \frac{L_{0,n}(P)}{L_{0,n}(Q)} = \inf_{\mathcal{Q}} \frac{L_{1,k}(P)}{L_{1,k}(Q)} \frac{L_{k,n}(P)}{A_{k,n}(P)} \frac{A_{k,n}(Q)}{L_{k,n}(Q)} \frac{A_{k,n}(P)}{A_{k,n}(Q)}$$

$$\ge \inf_{\mathcal{Q}} \left(\frac{L_{0,k}(P)}{L_{0,k}(Q)}\right) \frac{L_{k,n}(P)}{A_{k,n}(P)} \inf_{\mathcal{Q}} \frac{A_{k,n}(Q)}{L_{k,n}(Q)} \inf_{\mathcal{Q}} \frac{A_{k,n}(P)}{A_{k,n}(Q)} . \tag{9}$$

The first term on the right-hand side of (9) is positive for all k by the assumptions on the offspring distribution. Choose k so that the second and third terms are close to 1 for all n and Q. It remains to show that for this k as $n \to \infty$

$$\inf_{\mathcal{Q}} \frac{A_{k,n}(P)}{A_{k,n}(Q)} \to \infty \tag{10}$$

a.s. under **P**, to deduce that

$$\inf_{\mathcal{Q}} \frac{L_{0,n}(P)}{L_{0,n}(Q)} \to \infty \text{ a.s. } (\mathbf{P}) .$$

To verify (10) write

$$\frac{A_{k,n}(P)}{A_{k,n}(Q)} = \exp\left(\frac{1}{2} \sum_{k+1}^{n} \frac{(Z_j - mZ_{j-1})^2}{Z_{j-1}} \left(\frac{1}{\sigma_Q^2} - \frac{1}{\sigma_P^2}\right)\right)\left(\frac{\sigma_Q}{\sigma_P}\right)^{n-k} .$$

In order to minimize this quantity over $\mathcal{Q}$ we simply plug in the maximum likelihood estimator of σ_Q^2 and m. Since we are looking at a normal likelihood, the unrestricted mle of σ^2 is $\hat{\sigma}^2 = 1/(n-k)\,\Sigma_{j=k}^{n}\,(Z_j - \hat{m}_{k,n} Z_{j-1})^2/Z_{j-1}$ where $\hat{m}_{k,n} = (Y_n - Y_k)/(Y_{n-1} - Y_{k-1})$ (cf. also section 5) and satisfies $|\hat{\sigma}^2 - \sigma_P^2| < \delta/2$ for n large enough. Thus the mle $\hat{\sigma}_{\mathcal{Q}}$ of σ^2 within $\mathcal{Q}$ is obtained on the boundary, and is $\sigma_P^2 \pm \delta$. Hence

$$\begin{aligned} \inf_{\mathcal{Q}} \frac{A_{k,n}(P)}{A_{k,n}(Q)} &\geq \exp\left(\frac{n-k}{2}\,\hat{\sigma}^2\left(\frac{1}{\hat{\sigma}_{\mathcal{Q}}} - \frac{1}{\sigma_P^2}\right)\right)\left(1 \pm \frac{\delta}{\sigma_P^2}\right)^{n-k} \\ &= \exp\left((n-k)\left\{\log(1 \pm \delta/\sigma_P^2) + \frac{1}{2}\,\frac{\hat{\sigma}^2}{\sigma_P^2}\left(\frac{1}{1 \pm \delta} - 1\right)\right\}\right) . \end{aligned}$$

Since $\hat{\sigma}^2 \to \sigma_P^2$ a.s., the term in curly brackets will eventually become positive, and therefore $\inf_{\mathcal{Q}} A_{k,n}(P)/A_{k,n}(Q) \to \infty$ a.s. (**P**), concluding the proof. □

Remark: The restriction to finite support and all probabilities bounded below is needed to avoid distributions in $\mathcal{Q}$ that are trivial or have lattice span arbitrarily close to something other than 1. It may be possible to establish a similar result in a more general setting, but this requires a uniform local central limit theorem allowing for different lattice sizes. Such a result is not yet available. □

Example: Monte Carlo Results

In addition to theoretical investigation of the estimator, we carried out a small Monte Carlo study. It indicated that $\hat{\sigma}^2$ has smaller mean squared error that Dion's estimator in all the cases studied. We did not use the bias-corrected version of Dion's estimator. Only small sample sizes were investigated, since the computational burden becomes rather heavy for large sample sizes. The first set of results are for a Poisson offspring distribution with mean 1.1 and varying sample sizes. In each case 100 nonextinct paths were used. There is a tendency for $\hat{\sigma}^2$ to be smaller than $\tilde{\sigma}^2$, except at the smallest sample size. This tendency becomes more pronounced as n gets larger.

	$n=3$		$n=5$		$n=7$		$n=10$		$n=15$	
	$\tilde{\sigma}^2$	$\hat{\sigma}^2$	$\tilde{\sigma}^2$	$\hat{\sigma}^2$	$\tilde{\sigma}^2$	$\hat{\sigma}^2$	$\tilde{\sigma}^2$	$\hat{\sigma}^2$	$\tilde{\sigma}^2$	$\hat{\sigma}^2$
Mean	0.779	0.707	0.843	0.698	0.989	0.800	1.111	0.985	1.074	0.985
Variance	0.876	0.407	0.504	0.268	0.716	0.387	0.356	0.240	0.235	0.208
mse	0.979	0.561	0.569	0.429	0.728	0.477	0.357	0.253	0.235	0.221

The next set of Monte Carlo results compares a Poisson and a geometric offspring distribution, with about the same mean. Here the sample size n is 10, and we used 100 nonextinct paths for each estimate.

	Poisson(1.2)		Geometric(0.45)	
	$\tilde{\sigma}^2$	$\hat{\sigma}^2$	$\tilde{\sigma}^2$	$\hat{\sigma}^2$
Mean	1.292	1.101	3.152	2.416
Variance	0.388	0.267	3.396	1.980
mse	0.396	0.277	3.586	2.070

The Poisson computations generally converged more rapidly, with an average number of iterations of 30, and an sd of 49, whereas for the geometric case the mean number was 135 with an sd of 81. □

5. THE AGE

Suppose that we observe one or several generations of a branching process, which started from one individual. How long has the process been running? For a specific application, Thompson (1976) studied rare protein variants in South American Indians, restricting attention to those alleles occurring only in a single South American tribe: the Yanomama tribe of the Amazonas. The question in this setting is: given the number of replicates of a particular variant in a population, how long ago did the mutation first arise?

Stigler (1970) was the first to look at the problem of estimating the age of a BGW process. He assumed a known offspring distribution of fractional linear form (i.e., a geometric distribution with modified zero term; cf. Exercise 1.3). The likelihood from one observation, conditional on non-extinction is

$$L(n) = \frac{P_n^{(k)}(0)}{k!(1 - P_n(0))} .$$

In the fractional linear case we can write

$$P_n(s) = 1 - \frac{b_n}{1 - c_n} + \frac{b_n s}{1 - c_n s} ,$$

so $\mathbf{P}(Z_n = k) = b_n c_n^{k-1}$ and $P_n(0) = (1 - b_n)/(1 - c_n)$. Thus the likelihood is

$$L(n) = (1 - c_n)c_n^{k-1} ,$$

where $c_n = (m^n - 1)/(m^n - q)$. Differentiating the logarithm of L (treating n as a continuous variable) we get the likelihood equation

$$\left(\frac{k-1}{m^n - 1} - \frac{k}{m^n - q}\right)(m^n \log m) = 0 ,$$

so the mle is

$$\hat{n} = \frac{\log(k(1-q) - q)}{\log m} . \tag{11}$$

Notice that if $q = 0$ we get the method of moment estimator of n.

Proposition 3.5: Given $Z_n > 0$ and $m > 1$, $\hat{n} - n \overset{\text{a.s.}}{\to} \log[(1-q)W]/\log m$ as $n \to \infty$.

Proof: Write

$$\hat{n} - n = \log((1-q)m^{-n}Z_n + qm^{-n})/\log m . \qquad \square$$

We see from the proposition that $\hat{n}$ is not consistent, even in the rather peculiar sense (the parameter becoming infinitely large) stated there. However, if we invent an even more peculiar definition of near consistency, we may obtain a result. Call an estimator T of θ *α-consistent* if $\theta^{-\alpha}(T - \theta) \to 0$ as $\theta \to \infty$.

Corollary: $\hat{n}$ is α-consistent on the explosion set for all $\alpha > 0$.

In the fractional linear generating function case, W has an exponential distribution with mean $1/(1-q)$, so we can use Proposition 3.5 to compute asymptotic confidence bands for n. Stigler's results indicate that the convergence to the asymptotic distribution is fairly rapid.

Application: Age of a Mutation

The genetic example mentioned above centers on the occurrence of a rare protein, an albumin variant, local to the Yanomama tribe. In the 47 tribal villages, 875 replicates of the variant were found among adults. The modified geometric distribution fits genetic data from this tribe quite well. In this application, we can think of m as the mean rate of increase of the mutant type, representing a selective advantage or disadvantage in a stable population, when compared to the overall growth rate. In other words, the offspring mean for the mutation is $\mu(1+s)$, where μ is the population offspring mean, and s measures the selective advantage or disadvantage.

Equation (11) shows that $\hat{n}$ is a function of m, q, and the observation k. It seems reasonable to assume that $m > 1$. Again using population data it was seen that, to a good approximation, p_k/p_{k-1} was approximately 0.67. With this assumption, we can plot the likelihood surface as a function of m and n. Figure 3.3 has the maximum likelihood ridge in the middle, and outside contour lines corresponding to the maximum likelihood minus 2 and 4, respectively. From population data, an offspring mean of 1.02 seems to correspond to the average growth rate, and corresponds to a mutation with no selective advantage ($s = 0$). This yields $\hat{n} = 168$, corresponding to about 4000 years. Using the geometric distribution of W, we easily compute an asymptotic confidence band:

$$0.95 = \mathbf{P}\left\{\hat{n} - \frac{\log\log 40}{\log m} \le n \le \hat{n} - \frac{\log\log 1.0256}{\log m}\right\}.$$

For this particular example, we get the band $(102, 353)$, which appears genetically reasonable. This band corresponds to a selectively neutral allele. One can compute bands under various assumptions of selective effect on the offspring mean. For example, an outlined in Exercise 3, an assumption of

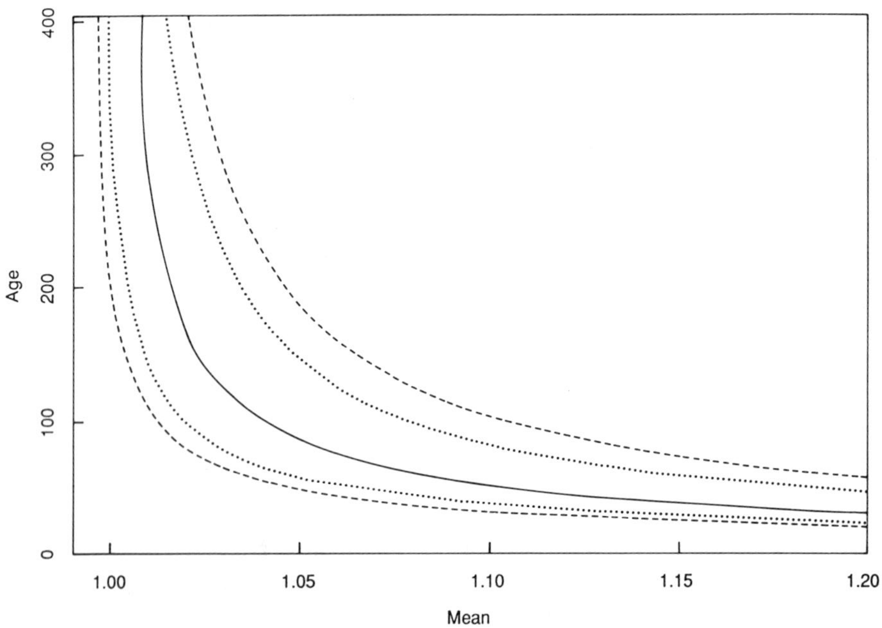

FIGURE 3.3. Likelihood surface for the Yanomana mutation age as a function of the offspring mean m and the age n. The solid line is the maximum ridge of the likelihood function, while the dotted line is 2 units of log likelihood below the maximum, and the dashed line is 4 units below.

$m = 1$, representing selective disadvantage ($s < 0$), yields $\hat{n} = 1313$, corresponding to about 30,000 years. This is probably longer than the period Indians have been in South America (although recent evidence suggests that it may be about the length of that period). Since the variant is local to one tribe, it would appear unlikely that it had arisen before tribal separation and disappeared in all tribes except one. Hence an offspring mean of 1 is an improbable choice. □

Crump and Howe (1972) study the case when m and q are estimated from data from several generations $Z_l, \ldots, Z_n$. Notice that since the Z_k form a Markov chain with stationary transition probabilities, the likelihood with respect to n is proportional to the marginal distribution of Z_l. Hence the mle of n is $\hat{l} + d$ where d is the number of observations taken (i.e., $n - 1$). The remainder of the data can be used to estimate the parameters. The nonparametric mle of m is

$$\hat{m}_{l,n} = \frac{\sum_{l+1}^{n} Z_k}{\sum_{l}^{n-1} Z_k} ,$$

which is consistent on A by the results in the last chapter if either $n \to \infty$ for fixed m or $m \to \infty$ for fixed d. The problem of estimating q is harder. For q near 1, the approximations in Exercise 1.7 can be used, e.g.,

$$\tilde{q} = \exp(-2(\hat{m}_{l,n} - 1)/\hat{\sigma}^2)$$

with $\hat{\sigma}$ the estimator from section 3.1, suitably modified. Another possibility, suggested by Crump and Howe, is to use the estimator of q resulting from a Poisson offspring distribution with mean $\hat{m}_{l,n}$. It turns out that the choice of estimator of q is fairly inessential in an asymptotic sense.

Proposition 3.6: Let $\tilde{n}$ be an estimator obtained by replacing m in (11) by $\bar{m}$ and q by any nonnegative statistic $q^* < 1$ a.s. on A. Then

$$\limsup_{n \to \infty} (\tilde{n} - n) < \infty \text{ a.s. on } A .$$

Proof

$$\tilde{n} - n = \frac{\log m^{-n} Z_n - n \log \dfrac{\bar{m}}{m} + \log(1 - q^*)}{\log \bar{m}} .$$

The only term that causes any trouble is $n \log(\bar{m}/m)$. Write this

$$\log\left(\frac{\bar{m}}{m}\right) + \log\left(\left(\frac{\bar{m}}{m}\right)^{\log Z_{n-1}}\right)^{(n-1)/(\log Z_{n-1})}$$

Since $Z_{n-1}^{1/(n-1)} \to m$ on A, we need only worry about $(\bar{m}/m)^{\log Z_{n-1}}$. From Theorem 2.3 we know that

$$Z_{n-1}^{1/2}(\bar{m}-m) \xrightarrow{d} \mathbf{N}(0, 1) .$$

Denote the random variable on the left-hand side by ξ_n. Then

$$\frac{\bar{m}}{m} = 1 + \frac{\xi_n}{mZ_{n-1}^{1/2}}$$

Thus

$$\log Z_{n-1} \log \frac{\bar{m}}{m} = \log Z_{n-1}\left(\frac{\xi_n}{mZ_{n-1}^{1/2}} + o_{\mathbf{P}}(Z_{n-1}^{-1/2})\right)$$

from which the statement of the theorem follows. □

It is possible to find the mle for some cases other than the fractional linear generating function. Generally, the problem is to find an explicit form for $P_n(s)$, or to develop an algorithm that allows for numerical optimization. Adès et al. (1982) prove the following unimodality result, which yields an algorithm for computing the mle for a general distribution with $p_0 = 0$.

Proposition 3.7: Suppose that $p_0 = 0$. If n is such that for a given k $\mathbf{P}(Z_{n+1} = j) < \mathbf{P}(Z_n = j)$ for all $j \le k$, then $\mathbf{P}(Z_{n+2} = j) < \mathbf{P}(Z_{n+1} = j)$ for all $j \le k$.

Proof: It suffices to look at a $k \ge 1$.

$$\begin{aligned}\mathbf{P}(Z_{n+2} = k) &= \sum_{j=1}^{k} \mathbf{P}(Z_{n+1} = j)\mathbf{P}(Z_{n+2} = k \mid Z_{n+1} = j) \\ &= \sum_{j=1}^{k} \mathbf{P}(Z_{n+1} = j)p(k\mid j) < \sum_{j=1}^{k} \mathbf{P}(Z_n = j)p(k\mid j) = \mathbf{P}(Z_{n+1} = k) .\end{aligned}$$

□

Adès et al. also prove a more general unimodality result which covers the Poisson case (first treated by Hwang and Hwang, 1978) as well as the geometric and the negative binomial offspring distributions.

6. THE EXTINCTION PROBABILITY

The extinction probability is an important parameter in many applications of branching processes. In using a BGW process to model (early stages of) an epidemic, the extinction probability corresponds to the probability of a minor outbreak. For a photon detector of multiplier type, the probability of successful detection of a photon is $1 - q$, a measure of the reliability of the device. We saw in the previous section that an estimator of q is needed to estimate the age of a process, although the quality of the estimator did not seem very important.

Unless q is a function of m and σ^2 only, we cannot estimate it consistently from observing just generation sizes. In this section we will look at two different aspects of the problem: first estimating q in our usual setting for a parametric model, and second nonparametric estimation when the entire family tree can be observed.

Application: Branched Polymers in Solution

The formation of branched polymers in solution occurs by molecules of the same type attaching to one molecule, which is called the repeat unit. If the repeat unit has f points of possible molecular attachment, all assumed equally reactive, it is called an f-functional. Attached molecules will be $(f-1)$-functional. The formation of polymers can be thought of as a branching process where the offspring distribution for the zeroth generation is different (Gordon, 1962; Good, 1963). The offspring distribution for the repeat unit is

$$P_0(s) = (1 - p + ps)^f$$

and for subsequent generations is

$$P_1(s) = (1 - p + ps)^{f-1} ,$$

where p is the probability of reaction at each functionality. The pgf for the total number of individuals, not counting the repeat unit, is

$$U(s) = sP_1(sP_1(s \cdots))$$

or

$$U(s) = sP_1(u(s)) . \tag{12}$$

The pgf U is defective if $P_1'(1) > 1$, since then there is positive probability of a tree being infinite. The pgf for the total size of finite trees is

$$W(s) = sF_0(U(s)) .$$

The soluble fraction $S(p)$ is the probability of finite polymers

$$S(p) = W(1) .$$

From (12) we see that $U(1) = P_1(U(1))$, so $U(1) = q^1$, the extinction probability for the offspring distribution P_1. Hence the soluble fraction can be written

$$S(p) = F_0(q^1) . \qquad \square$$

In the parametric case we have an underlying offspring distribution with pgf $P(s) = P(s; \boldsymbol{\theta})$ for $\boldsymbol{\theta} \in \Theta \subset \mathbf{R}^d$. We will assume that we can estimate $\boldsymbol{\theta}$ consistently using the estimator $\tilde{\boldsymbol{\theta}}$ (so if we only observe generation sizes and the parameters are variation independent we must have $d \leq 2$). The natural estimator for q is

$$\tilde{q} = \inf\{s: P(s, \tilde{\boldsymbol{\theta}}) = s\} .$$

Theorem 3.4: Let $\boldsymbol{\theta}^0$ be the true parameter value. If

(i) $m(\boldsymbol{\theta}^0) > 1$, $(\partial/\partial s)P(s; \boldsymbol{\theta})$ is jointly continuous in s and $\boldsymbol{\theta}$, and the quantities $a_i = (\partial/\partial\theta_i)P(q; \boldsymbol{\theta}^0)$ exist,

(ii) $Y_{n-1}^{1/2}(\tilde{\boldsymbol{\theta}} - \boldsymbol{\theta}^0) \xrightarrow{d} \mathbf{N}(0, \Sigma)$ with $\mathbf{a}^T\Sigma\mathbf{a} > 0$,

then

$$Y_{n-1}^{1/2}(\tilde{q} - q) \xrightarrow{d} \mathbf{N}(0, h^{-2}\mathbf{a}^T\Sigma\mathbf{a})$$

where $h = 1 - (\partial/\partial s)P'(q; \boldsymbol{\theta}^0)$.

Proof: First note that $\tilde{\boldsymbol{\theta}} \to \boldsymbol{\theta}^0$ in probability by (ii). Thus $P(s; \tilde{\boldsymbol{\theta}}) \to P(s; \boldsymbol{\theta})$. By continuity the convergence is uniform in $s \in [0, 1]$. The same holds for $(\partial/\partial s)P$. Expand

$$\begin{aligned} 0 &= P(\tilde{q}, \tilde{\boldsymbol{\theta}}) - \tilde{q} \\ &= P(q; \tilde{\boldsymbol{\theta}}) - q + (\tilde{q} - q)\left[\frac{\partial}{\partial s} P(s_n; \tilde{\boldsymbol{\theta}}) - 1\right], \end{aligned}$$

where $|s_n - q| \leq |\tilde{q} - q|$. Let $h_n = 1 - (\partial/\partial s)P(s_n; \tilde{\boldsymbol{\theta}})$. By the uniform convergence of $P(s; \tilde{\boldsymbol{\theta}})$ to $P(s; \boldsymbol{\theta}^0)$ we see that $\tilde{q} \to q$ in probability, and by the uniform convergence of $(\partial/\partial s)P$ we see that h_n converges in probability to $h = 1 - (\partial/\partial s)P(q; \boldsymbol{\theta}^0)$. Clearly $0 \leq h \leq 1$. If $h = 0$ then the function $f(s) = (\partial/\partial s)P(s; \boldsymbol{\theta}^0) - 1$ must vanish both at $s = q$ and at a point in the open interval $(q, 1)$. The latter is because $P(s) - s$ vanishes at q and 1, so its derivative $f(s)$ must have a zero in that open interval (Rolle's theorem). Hence there must be a point in $(q, 1)$ such that $f''(s) = 0$. But since P is a power series, this means that $p_0 + p_1 = 1$, contradicting that $m > 1$. Hence

$h > 0$. We now write

$$h_n(\tilde{q} - q) = P(q; \tilde{\boldsymbol{\theta}}) - q$$
$$= P(q; \boldsymbol{\theta}^0) - q + \sum_{i=1}^{d} (\tilde{\theta}_i - \theta_i) \frac{\partial}{\partial \theta_i} P(q; \mathbf{r}_{ni})$$

where $\mathbf{r}_{ni} \to \boldsymbol{\theta}^0$ as $n \to \infty$ for each i. It follows that

$$Y_{n-1}^{1/2}(\tilde{q} - q) = h_n^{-1} \sum_{i=1}^{d} Y_{n-1}^{1/2}(\tilde{\theta}_i - \theta_i^0) \frac{\partial}{\partial \theta_i} P(q; \mathbf{r}_{ni})$$

from which the claim follows, using (ii). □

Example: Poisson and Geometric Distributions

If the underlying distribution is misspecified, the estimate $\tilde{q}$ may be quite wrong. Figure 3.4 shows the true q for the Poisson and geometric offspring distributions. For small values of the offspring mean the two values of q are similar, but the discrepancy grows rapidly with the mean. Suppose that

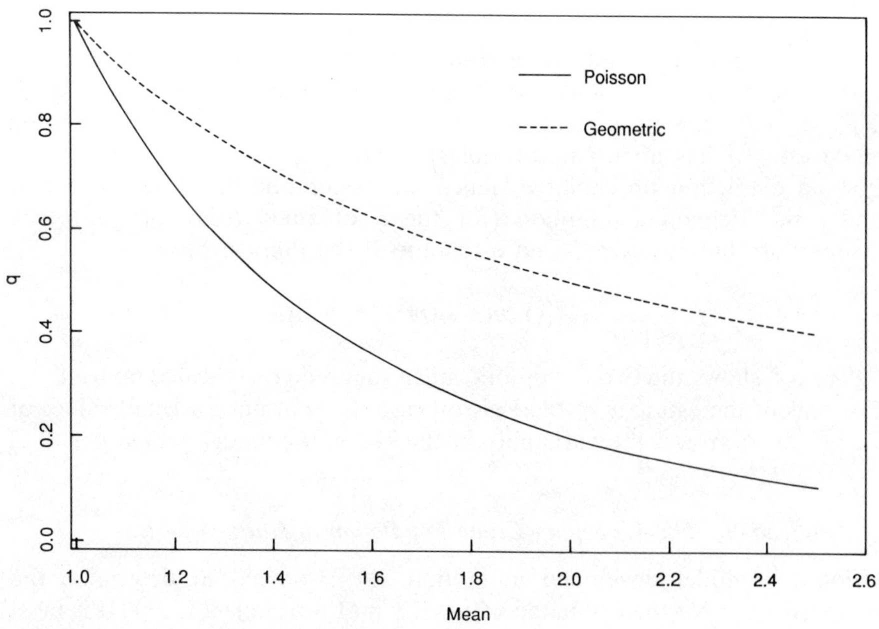

FIGURE 3.4. The extinction probability q as a function of m for the Poisson (solid) and geometric (dashed) offspring distributions.

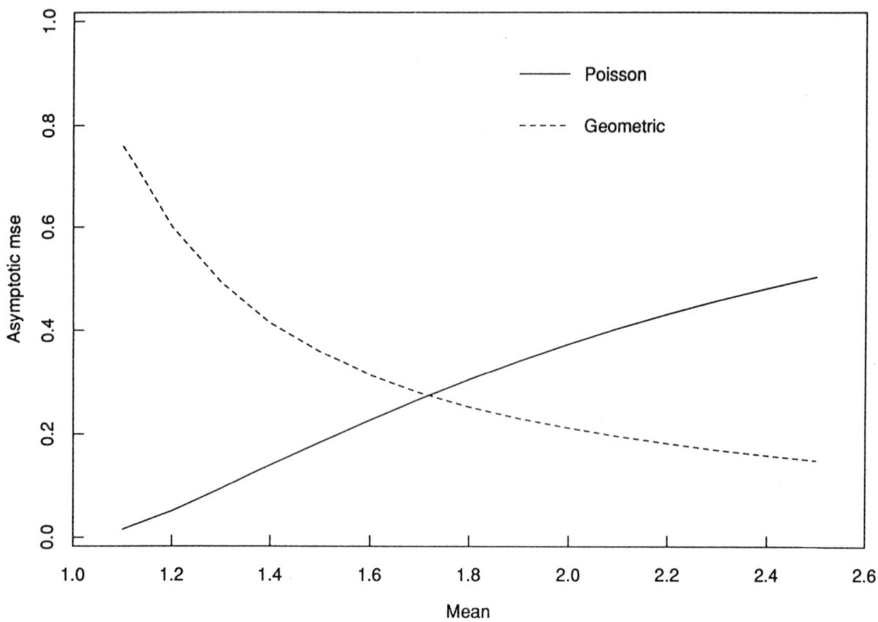

FIGURE 3.5. Mean squared error when the Poisson (solid) and geometric (dashed) distribution-based $\tilde{q}$ is used to estimate q for a Poisson offspring distribution as a function of the offspring mean.

we believe that the underlying distribution is geometric. Then $\tilde{q} = 1/\hat{m}$, and Theorem 3.3 yields that the asymptotic variance of $\tilde{q}$ is $Y_{n-1}^{-1}(m-1)^2(m+1)/m^3$. On the other hand, if the underlying distribution is Poisson, $\tilde{q}$ has an asymptotic bias of $1/m - q(m)$, where $q(m)$ is the Poisson extinction probability. Hence the asymptotic mean squared error under the Poisson assumption (for the $\tilde{q}$ obtained from the geometric assumption, but regarded as an estimator of the Poisson q) is

$$Y_{n-1}^{-1}[(1/m - q(m))^2 + m^{-3}].$$

Figure 3.5 shows the two asymptotic mean squared errors scaled up by Y_{n-1}. The bias of the estimate in the Poisson case is important for small values of m, but as m grows, the variability in the geometric model gets larger. □

Application: The Whooping Crane Population of North America

Using the model developed in section 2.4, we need to determine the standard error for the estimated value of $q = (1-b-c)/c(1-c)$. It is most convenient to use the parametrization in terms of p_0 and c. Then the likelihood becomes

$$p_0^{N_0}[(1-p_0)(1-c)]^{Y_{n-1}-N_0} c^{Y_n - z_0}$$

whence the asymptotic variance–covariance matrix of $(\hat{p}_0, \hat{c})$ is

$$\Sigma = \begin{pmatrix} \dfrac{p_0^2(1-p_0)^2}{N_0(1-2p_0)+Y_{n-1}p_0^2} & 0 \\ 0 & \dfrac{c^2(1-c)^2}{(Y_n-1-N_0)c^2-(Y_n-z_0)(1-c)^2} \end{pmatrix}.$$

It now follows from Theorem 3.3 that the asymptotic variance of $\hat{q}$ is

$$\frac{(1-p_0)^2}{Y_{n-1}(c-p_0)^2}\left(\frac{(c-p_0)^2}{c^2(1-p_0)^2}\,\Sigma_{11}+\frac{p_0^2(p_0-c)^2}{(1-p_0)^2c^4}\,\Sigma_{22}\right).$$

Using the estimated N_0 to compute the asymptotic variance we see that $\hat{q}=0.917$ with a standard error of 0.00110. A more useful quantity is the estimate of the probability of extinction of the population, based on the last observed value (51). This is $\hat{q}^{51}$, with an estimated asymptotic standard error of $51\hat{q}^{50}\,\mathrm{se}(\hat{q})$. This yields an approximate confidence interval for the probability of extinction for the 1972 population of (0.0106, 0.0135). Keiding (1976) argues that this estimate is rather too optimistic (cf. also Exercise 5.3). □

Now to the nonparametric case. Assume that we can observe $n_{l,k}=\Sigma_{i=1}^{Z_{l-1}}1(X_{l,i}=k)$, and let $N_k(n)=\Sigma_{l=1}^{n}n_{l,k}$. As in section 2.4 we let $\hat{p}_k=N_k(n)/Y_{n-1}$. For a given k we introduce the $\mathscr{F}_n^X$-martingale $M_n=N_k(n)-p_kY_{n-1}$. Since given $\mathscr{F}_{n-1}^X$, $n_{n,k}\sim\mathrm{Bin}(Z_{n-1},p_k)$ we have

$$\begin{aligned}\mathbf{E}(M_n|\mathscr{F}_{n-1}^X)&=\mathbf{E}(N_k(n-1)+n_{n,k}-p_kY_{n-1}|\mathscr{F}_{n-1}^X)\\&=N_k(n-1)+p_kZ_{n-1}-p_kY_{n-1}=M_{n-1}\end{aligned}$$

verifying the martingale property. This martingale will be useful in determining properties of $\hat{p}_k$.

Proposition 3.8

(i) $\hat{p}_k\to p$ a.s. on A.

(ii) $$Y_{n-1}^{1/2}\begin{pmatrix}\hat{p}_0-p_0\\ \cdots\\ \hat{p}_K-p_K\end{pmatrix}\xrightarrow{d}\mathbf{N}\left(\begin{pmatrix}0\\ \cdots\\ 0\end{pmatrix},\begin{pmatrix}p_0(1-p_0)&\cdots&-p_0p_K\\ \cdots&\cdots&\cdots\\ -p_0p_K&\cdots&p_K(1-p_K)\end{pmatrix}\right)$$

on A for any K.

Proof: (i) By the martingale law of large numbers (Lemma 2.5), $Y_{n-1}^{-1}M_n\to 0$ a.s. on the set

$$\left\{Y_{n-1}\to\infty,\ \sum_{j=1}^{n}Y_{n-1}^{-2}\mathbf{E}(D_j^2|\mathscr{F}_{j-1}^X)<\infty\right\},$$

where $D_j = M_j - M_{j-1} = n_{j,k} - p_k Z_{j-1}$. Compute

$$\mathbf{E}(D_j^2 | \mathscr{F}_{j-1}^X) = p_k(1-p_k)Z_{j-1}$$

so that

$$\sum_{j=1}^{n} Y_{n-1}^{-2} \mathbf{E}(D_j^2 | \mathscr{F}_{j-1}^X) = p_k(1-p_k) \sum_{j=1}^{n} Y_{n-1}^{-2} Z_{j-1}$$

which is always finite by Lemma 2.6.

To prove (ii) we will use Proposition 2.7 together with the Cramér–Wold device. Let $a_0, \ldots, a_K$ be an arbitrary sequence of real numbers, and let $D_j = \Sigma_{i=0}^{K} a_i(n_{j,i} - p_i Z_{j-1})$. Then

$$V_n^2 = \sum_{j=1}^{n} \mathbf{E}(D_j^2 | \mathscr{F}_{j-1}^X) = Y_{n-1}\Big(\sum a_i^2 p_i(1-p_i) - \sum_{i \neq j} a_i a_j p_i p_j\Big)$$

and

$$\begin{aligned}\mathbf{Var}\, M_n &= \mathbf{E}(\mathbf{Var}(M_n | \mathscr{F}_{n-1}^X)) + \mathbf{Var}(\mathbf{E}(M_n | \mathscr{F}_{n-1}^X) \\ &= \mathbf{E} Z_{n-1}\Big(\sum a_i^2 p_i(1-p_i) - \sum_{i \neq j} a_i a_j p_i p_j\Big) + \mathbf{Var}\, M_{n-1}\,,\end{aligned}$$

so that

$$s_n^2 = \mathbf{Var}\, M_n = \frac{m^n - 1}{m-1}\Big(\sum a_i^2 p_i(1-p_i) - \sum_{i \neq j} a_i a_j p_i p_j\Big)\,.$$

Thus $s_n^{-2} V_n^2 \rightarrow W$ a.s. on A, and $s_{n-r}^2 / s_n^2 = m^{-r}$. It remains to check that

$$\mathbf{E}(\exp(itD_n / aV_n) | \mathscr{F}_{n-1}^X) \rightarrow \exp(-t^2/2)$$

on $\{W > 0\} = A$. But $(n_n, 0, \ldots, n_{n,K})^T \sim \mathrm{Mult}(Z_{n-1}, (p_0, \ldots, p_K)^T)$, so

$$\frac{D_n}{aV_n} = \frac{\sum a_i(n_{n,i} - p_i)Z_{n-1}}{\Big(Z_{n-1}\Big(\sum a_i^2 p_i(1-p_i) - \sum_{i \neq j} a_i a_j p_i p_j\Big)\Big)^{1/2}} \left(\frac{Z_{n-1}}{aY_{n-1}}\right)^{1/2}.$$

On A, the first term on the right-hand side is asymptotically standard normal, and the second converges to 1 provided that we take $a = m/(m-1)$. Applying Proposition 2.7 we see that

$$\sum_{1}^{n} \frac{D_j}{\frac{m-1}{m} V_n} \xrightarrow{d} \mathbf{N}\Big(0, \frac{m}{m-1}\Big)\,,$$

from which the claim follows. □

Now define $P_n(s) = \Sigma_{j=0}^{\infty} \hat{p}_j s^j$. This is the nonparametric maximum likelihood estimator of the pgf $P(s)$ based on observing the family tree.

Theorem 3.5: Conditional upon nonextinction,

(i) $P_n(s) \to P(s)$ uniformly in s.
(ii) $(Y_{n-1}(P(s^2) - P^2(s)))^{-1/2}(P_n(s) - P(s)) \xrightarrow{d} \mathbf{N}(0, 1)$.

Proof: To prove (i), estimate

$$|P_n(s) - P(s)| = \left| \sum_{j=0}^{K} (\hat{p}_j - p_j) + \sum_{j=K+1}^{\infty} (\hat{p}_j - p_j)s^j \right|$$
$$\leq \max_{0 \leq j \leq K} |\hat{p}_j - p_j| + \sum_{j=K+1}^{\infty} s^j .$$

Now take K large so that the second term is small (uniformly in s since the bound is a limit of monotone functions on a compact set). Then take n large and use Proposition 3.10(i) to make the first term small (this does not involve s).

To verify (ii) we use Proposition 2.7 again. The martingale is

$$M_n = \sum s^k(N_k(n) - Y_{n-1}p_k) .$$

Then

$$\mathbf{E}(D_n^2 | \mathcal{F}_{n-1}^X) = Z_{n-1}(P(s^2) - P^2(s)) .$$

The argument now is very similar to that of Proposition 3.10. We have $s_n^{-2}V_n \to W$, and $s_n^{-2}s_{n-r}^2 = m^{-r} + o(1)$. It remains to show that

$$\mathbf{E}(\exp(itD_n/aV_n) | \mathcal{F}_{n-1}^X) \to \exp(-t^2/2)1(W > 0) .$$

Write $D_n^K = \Sigma_0^K s^k(n_{n,k} - Z_{n-1}p_k)$. By the argument in Proposition 3.10 this is asymptotically normal. We must show that the remainder $D_n - D_n^K$ can be made negligible compared to V_n. Compute

$$\mathbf{Var} \sum_{K+1}^{\infty} Z_{n-1}^{-1/2}(n_{k,n} - Z_{n-1}p_k)s^k | \mathcal{F}_{n-1}^X) \leq \sum_{K+1}^{\infty} p_k(1 - p_k)s^k \leq \sum_{K+1}^{\infty} s^k$$

which can be made arbitrarily small. By Chebyshev's inequality the normalized remainder is asymptotically negligible. □

Now we are ready to estimate q. Let $\hat{q} = \inf\{s\colon P_n(s) = s\}$. This is the nonparametric mle based on observing the family tree.

Theorem 3.6

(i) $\hat{q} \to q$ a.s. on A.
(ii) Provided that $m > 1$, $Y_{n-1}^{1/2}(\hat{q} - q) \xrightarrow{d} \mathbf{N}(0, \tau^2)$, where $\tau^2 = (m-1)(P(q^2) - q^2)(1 - P'(q))^{-1}$.

Proof: Part (i) is immediate from Theorem 3.4(i). For part (ii), write

$$P_n(q) = P_n(\hat{q}) + (q - \hat{q})P_n'(r_n) = \hat{q} - (\hat{q} - q)P_n'(r_n),$$

where $|r_n - 1| \le |\hat{q} - q|$ and $r_n \in \mathscr{F}_n^X$. Thus

$$Y_{n-1}^{1/2}(P_n(q) - P(q)) = Y_{n-1}^{1/2}(\hat{q} - q)(1 - P_n'(r_n)).$$

The left-hand side converges in distribution to $\mathbf{N}(0, (m-1)(P(q^2) - q^2))$ and $P_n'(r_n) \to P'(q)$ a.s., using Exercise 7. □

Example: Asymptotic Efficiency of the Nonparametric Estimator

In the Poisson case, where $\theta = m$, we can compare the asymptotic variance of the parametric and nonparametric estimators. Let $\tilde{\theta} = \hat{m}$, so that $Y_{n-1}^{1/2}(\tilde{\theta} - \theta^0) \xrightarrow{d} \mathbf{N}(0, \theta^0)$. Since $P(s; \theta^0) = \exp(\theta^0(s-1))$ we see that $(\partial/\partial s) P(q, \theta^0) = q\theta^0$, and $(\partial/\partial\theta)P(q, \theta^0) = (q-1)q$. Hence the asymptotic variance of $\tilde{q}$ is $\theta^0(q(q-1)/(1 - \theta^0 q))^2$ and that of $\hat{q}$ is $(\exp(\theta^0(q^2 - 1)) - q^2)(1 - \theta^0 q)^{-2}$. The asymptotic relative efficiency of the nonparametric mle $\hat{q}$ relative to the parameteric mle $\tilde{q}$ theefore is

$$e(\theta) = \frac{\theta(q(q-1))^2}{\exp(\theta(q^2 - 1)) - q^2}.$$

As outlined in Exercise 8, $e(\theta) \to 1$ as $\theta \to 1+$ and $e(\theta) \to 0$ as $\theta \to \infty$. Figure 3.6 is a graph of $e(\theta)$. We lose a lot of efficiency for large values of θ. On the other hand, if only generation sizes are observed we cannot compute $\hat{q}$. For small values of θ the efficiency loss is not large, and nonparametric mle has a built-in defense against misspecification of the offspring distribution.

7. SOURCES

The electron multiplier has always been a mainstay example of a branching process. Variance estimation was studied by Heyde (1974) and Dion (1975).

Lemma 3.3 was first proved by Dvoretzky and Wolfowitz (1951). The results on lattice size estimation were developed jointly with Richard Lockhart.

Dion et al. (1982) produced the first computable nonparametric mle for the offspring distribution. The use of the EM-algorithm for this problem appears new.

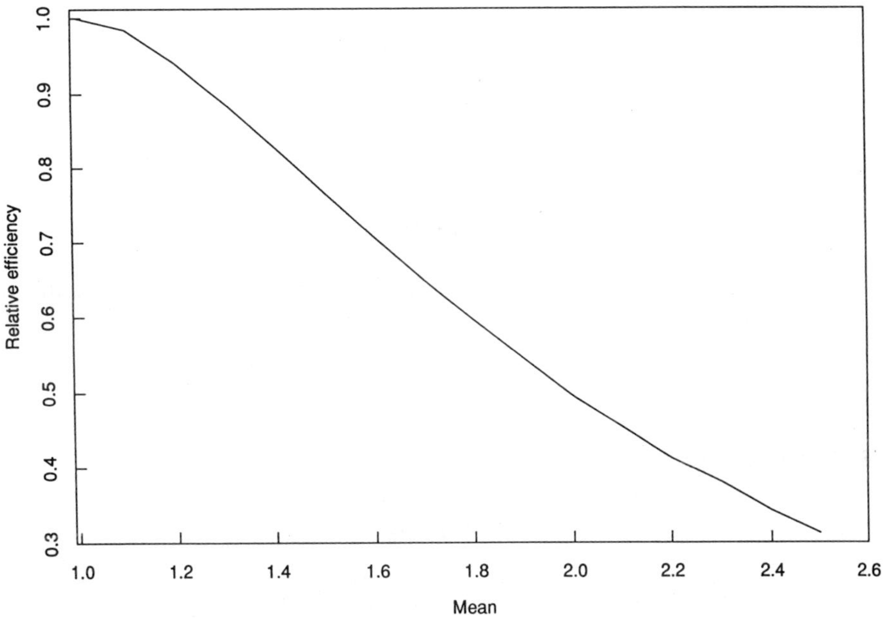

FIGURE 3.6. Asymptotic relative efficiency of the nonparametric mle to the parametric mle in the Poisson case.

The consistency of the mle of the offspring variance was proved jointly with Richard Lockhart.

Stigler (1970) first estimated the age of a branching process. His work was further developed by Crump and Howe (1972) and Adès et al. (1982). The application follows Thompson (1976).

Stigler (1971) studied the estimation of q from a sample from the offspring distribution (rather than from the process). Cryer and Robertson (1975) looked at estimation of q under the assumption of a unimodal offspring distribution, and derived some results also under the assumption that a sample from the offspring distribution was available. Pakes (1975) found properties of the nonparametric estimator of q from observing all offspring in the nth generation. Nanthi (1983, Theorem 4.5) derived the behavior of the nonparametric estimator based on the entire family tree up to generation n. The application to branching polymers is based on work by Good (1963) and Gordon (1962).

EXERCISES

1. Verify the variance formula (1).

2. Show that in the algorithm for nonparametric estimation of $(\hat{p}_j)$, each iteration has mean $\Sigma\, j\hat{p}_j^{(l)} = \hat{m}_n$.

3. Suppose that we have a geometric offspring distribution with modified zero term and mean $m=1$. Parametrize the distribution so that $p_0 = 1 - h/(1+h)$, $p_k = h^2(1+h)^{-(k+1)}$, $k \geq 1$.

(a) Show that

$$\mathbf{E}(Z_n | Z_n \neq 0) = (h+n)/h \,,$$
$$\mathbf{Var}(Z_n | Z_n \neq 0) = n(h+n)/h^2 \,.$$

(b) Show that the mle of n for known h is $\hat{n} = h(Z_n - 1)$ and deduce that $\hat{n}$ is conditionally unbiased with variance $n(n+h)$.

(c) Using the data in the Application of section 4, estimate n and compute a standard error for the estimate.

4. Prove that for all k with $p_k > 0$

$$\sum_{j \neq k} \mathbf{Cov}(\hat{p}_j, \hat{p}_k) < 0 \,.$$

5. (a) For the case of a Poisson offspring distribution show that

$$\hat{p}_j = \frac{\mathbf{E}(n_j | \mathcal{F}_n)}{Y_{n-1}} = \frac{1}{Y_{n-1}} \sum_{m=1}^{n} Z_{k-1}^{1-j} \left(1 - \frac{1}{Z_{k-1}}\right)^{Z_k - j} \binom{Z_k}{j} \,.$$

(b) Compare $\hat{p}_j$ to the mle of the Poisson offspring probabilities, namely $p_j(\hat{m})$.

6. (a) Prove that $P_n^{(k)}(s) \to P^{(k)}(s)$ a.s. as $n \to \infty$, uniformly for $0 \leq s \leq s_1 < 1$ for any $k \geq 1$.

(b) If $P^{(k)}(1) < 1$, the result in (a) holds for $s_1 = 1$.

(c) If $p_0 > 0$ and $p_0 + \cdots + p_{k-1} < 1$, then

$$\frac{P_n^{(k)}(s) - P^{(k)}(s)}{\mathbf{Var}^{1/2}\, P_n^{(k)}(s)} \xrightarrow{d} \mathbf{N}(0, 1) \,.$$

7. Show that $P_n'(\hat{q}) \to P'(q)$ a.s. as $n \to \infty$.

8. Let $e(\theta)$ be the asymptotic relative efficiency of $\hat{q}$ relative to $\tilde{q}$ for a Poisson offspring distribution, as computed in the second Example in section 3.5.

(a) Prove that $e(\theta) \to 1$ as $\theta \to 1+$.
Hint: Taylor expand $P(s; \theta)$ as a function of s about q. Then show that $(1 - q\theta)/(1-q) \to \frac{1}{2}$.

(b) Prove that $e(\theta) \to 0$ as $\theta \to \infty$.
Hint: Show that $q \leq e^{-\theta}(1 - \theta q)^{-1}$.

9. Compute the asymptotic efficiency of $\tilde{q}$ relative to $\hat{q}$ for a geometric offspring distribution.

10. (a) If $m < 1$ show that $\hat{q}$, the nonparametric mle from the whole family tree, is strongly consistent.

(b) If $m = 1$ show that

$$\mathbf{P}(Y_{n-1}^{1/2}(1 - \hat{q}) \leq x) \rightarrow \Phi(\tfrac{1}{2}x[P''(1)]^{1/2})1(x \geq 0) .$$

11. For the data on Hutterites in the Appendix, test the hypothesis that the distributions of male offspring and of female offspring are the same.

CHAPTER 4

Bayes Methods

We will now completely change perspective, and regard the parameter of interest as a random variable. One can then pose questions such as "What is the probability that the offspring mean is greater than one?" Questions of consistency become less important, and since this paradigm requires conditioning on the observations, no doubt about which is the appropriate conditional inference needs to arise. The price to be paid is that there frequently are computational problems with the Bayes approach. Some approximation methods will be discussed, and we also present a nonparametric approach to Bayes inference in branching processes.

1. INTRODUCTION

In the Bayesian statistical approach (see, e.g., Jeffreys, 1983), an unknown parameter θ is assumed to have a distribution $f(\theta)$ a priori. This prior distribution can be obtained in many ways, although the most common is to choose $f(\theta)$ among a convenient class of smooth densities with location and spread determined by the experimenters expectation and uncertainty about this expectation, respectively. An experiment is carried out to obtain further information about θ. A posteriori, that is after a datum x has become available, the distribution of θ is updated to the posterior (conditional) density of θ, given x. Usually this update is carried out using Bayes' formula

$$f(\theta|x) = \frac{f(\theta)f(x|\theta)}{\int f(\theta)f(x|\theta)\,d\theta},$$

where $f(x|\theta)$ is the distribution of the data, given the parameter (in other words the usual model of the data). Bayes' formula is a convenient way of updating one's current opinion about the parameter θ, and the posterior density reflects the opinion about θ in the light of the observation x. However, this is only one possible interpretation of the quantities involved.

A frequency interpretation requires obtaining $f(\theta)$ from previous experiments. This does not affect the mathematics involved, just the interpretation of the mathematics.

The idea of conjugate prior distributions was made popular by Raiffa and Schleiffer (1961). A family of priors is conjugate to a model if the posterior distribution, given a datum, is in the same family as the prior distribution. An ideal conjugate family is flexible enough to describe different prior expectations, yet small enough to be easy to handle mathematically.

Example: The Poisson Distribution and a Conjugate Prior

Let $f(x|\theta) = \theta^x e^{-\theta}/x!$ be the Poisson model. One way to find conjugate prior distributions, is to look at the model density as a function of the parameter, and see if this has familiar form. In this case we are lead to a gamma prior, $f(\theta) = \theta^{a-1} b^a e^{-b\theta}/\Gamma(a)$. Using Bayes' formula we see that the posterior is proportional to $\theta^{a+x-1} e^{-(b+1)\theta}$, whence θ a posteriori has the gamma distribution with parameter $a + x$ and $b + 1$. The gamma family can take on several different shapes, and therefore can describe different prior expectations. Whether it is flexible enough for a particular use has to be judged by the experimenter.

A huge class of prior distributions can be generated by letting $h(\theta)$ be proportional to an arbitrary probability density. If we use a prior proportional to $h(\theta)\theta^{a-1} \exp(-b\theta)$, the posterior density after observing x is proportional to $h(\theta)\theta^{a+x-1} \exp(-(b+1)\theta)$, which is of the same form as the prior. Hence *any* density can be used as a prior, and is formally a conjugate prior. □

In order to find a reasonably small class of conjugate priors, Diaconis and Ylvisaker (1979) suggest looking at priors with linear posterior expectation. Suppose that we have a exponential family model, so

$$f(x|\theta) = \exp(x\theta - M(\theta)), \qquad x \in \mathbf{X} \subset \mathbf{R},$$

where $M(\theta) = \log \int \exp(x\theta)\, dx$ (in the discrete case the integral is read as a sum) and $\theta \in \Theta = \{\theta: M(\theta) < \infty\}$. The set Θ is open and convex. We choose a family of conjugate priors π_{n_0, x_0} with density proportional to $\exp(n_0(x_0\theta - M(\theta)))$. Such priors exist at least if $n_0 > 0$ and $x_0 \in \mathbf{X}$.

Example: Geometric Distribution

Here $f(x|\theta) = \exp(x\theta - \log(1/(1 - e^\theta)))$, where $1 - e^\theta$ is the success probability. This is the natural parametrization of the geometric distribution. The natural parameter space is $\Theta = (-\infty, 0)$ and $\mathbf{X} = \mathbb{N}$. The conjugate family has density proportional to $(1 - e^\theta)^{n_0} e^{n_0 x_0 \theta}$. In terms of the success probability, this corresponds to a beta prior with parameters $(n_0+1,\ n_0x_0)$. The prior can also be defined for $-1<n_0<0$, provided $x_0<0$, but the resulting Bayes estimator (for squared error loss) has infinite Bayes risk (see below for a discussion of these concepts). □

For this family of conjugate priors we can prove the following result.

Proposition 4.1: If $\theta \sim \pi_{n_0,x_0}$, $n_0>0$, $x_0 \in \mathbf{X}$, then $\mathbf{E}_\theta \mathbf{E}(X|\theta)=x_0$.

Proof: Let $\Theta=(\underline{\theta}, \bar{\theta})$. By exponential family theory $\mathbf{E}(X|\theta)=M'(\theta)$. Since

$$\frac{d}{d\theta}\, e^{n_0(x_0\theta - M(\theta))} = n_0x_0e^{n_0(x_0\theta - M(\theta))} - n_0M'(\theta)e^{n_0(x_0\theta - M(\theta))}\,,$$

we can integrate by parts to obtain

$$\int M'(\theta)e^{n_0(x_0\theta - M(\theta))}\,d\theta = -\frac{e^{n_0(x_0\theta - M(\theta))}}{n_0}\Bigg|_{\underline{\theta}}^{\bar{\theta}} + x_0\int e^{n_0(x_0\theta - M(\theta))}\,d\theta\,.$$

The first term on the right-hand side is zero. To see this, first assume that $\bar{\theta}=\infty$. Since the function is proportional to density, it must go to zero as $\theta \to \infty$. If $\bar{\theta}<\infty$, let $\theta^* \in \Theta$, so

$$\int_{x\le 0} e^{\bar{\theta}x}\,dx \le \int_{x\le 0} e^{\theta^* x}\,dx < \infty\,.$$

Since $M(\bar{\theta})=\infty$ by construction, we have by monotone convergence that

$$\int_{x>0} e^{\theta x}\,dx \to \int_{x>0} e^{\bar{\theta}x}\,dx = \infty$$

as $\theta^* \le \theta \to \bar{\theta}$. Hence $M(\theta)\to\infty$ as $\theta \to \bar{\theta}$. As similar argument works for $\underline{\theta}$. □

Corollary: The posterior expectation after observing $X=x$ is

$$\mathbf{E}(\theta|X=x) = x_0 + \frac{x-x_0}{n_0+1}$$

which is linear in x. This suggests interpreting the prior parameters as equivalent to n_0 observations with a total of x_0.

Proposition 4.1 has a partial converse: If $\Theta=(-\infty, \bar{\theta})$, $\bar{\theta}<\infty$, the prior is not trivial, and $\mathbf{E}(M'(\theta)|X=x)=ax+b$, then $a>0$ and the prior has density proportional to $\exp((b\theta-(1-a)M(\theta))/a)$.

Example: Geometric Distribution

In the geometric case $\bar{\theta} = 0$, so any nontrivial prior with linear posterior expectation has to be a beta prior for $p = 1 - e^{\theta}$ as derived above. □

The posterior distribution contains all the currently available information about θ, and is a complete description of our knowledge. For example, from the posterior one can construct a credibility set, i.e., a set of probable values of θ. The most common method is to look at points of high posterior density (HPD credibility set), namely $A(c) = \{\theta: f(\theta|x) \geq c\}$. We choose c so that $\int_{A(c)} f(\theta|x)\, dx = 1 - \alpha$ for a credibility coefficient of $1 - \alpha$. If the posterior density is unimodal, the credibility set is an interval, and has the shortest length of all credibility intervals (Exercise 4).

Example: Geometric Distribution

Let the prior be a $B(\alpha, \beta)$ in terms of the success probability. The posterior after observing $x_1, \ldots, x_n$ is $B(\alpha + n, \beta + \Sigma\, x_i)$, which is a unimodal density. We can determine a 95% HPD interval for p by numerical methods. If n is large, the posterior distribution is approximately normal, so the HPD interval is approximately

$$\frac{\alpha + n}{\alpha + \beta + n + \sum x_i} \pm 1.96 \left(\frac{(\alpha + n)\left(\beta + \sum x_i\right)}{\alpha + \beta + n + 1 + \sum x_i} \right)^{1/2} \frac{1}{\alpha + \beta + n + \sum x_i} .$$

□

If we need a point estimate of θ, any location estimate based on the density $f(\theta|x)$ can be used. In order to choose a particular one, the tools of decision theory are frequently used. Let the loss function $L(\theta, \tilde{\theta})$ describe the loss (in some appropriate unit) incurred when estimating θ by $\tilde{\theta}$, i.e., when describing the distribution of θ using the number $\tilde{\theta}$. The Bayes risk of using $\tilde{\theta}$ is $r(\tilde{\theta}) = \mathbf{E}_{\theta} L(\theta, \tilde{\theta})$. The best estimator is that which minimizes the Bayes risk; it is called a Bayes estimator or a Bayes rule. Notice that we do not require that any data be used in order to estimate θ. However, if we have data we can use repeated conditioning to see that

$$\mathbf{E}_{\theta} L(\theta, \tilde{\theta}) = \mathbf{E}_X \mathbf{E}_{\theta|X} L(\theta, \tilde{\theta}) .$$

This identity sometimes can be used to find a Bayes rule. Suppose that we can find a $\tilde{\theta} = \tilde{\theta}(X)$ that minimizes $\mathbf{E}_{\theta|X} L(\theta, \tilde{\theta})$ for each X. Then this $\tilde{\theta}$ must be a Bayes rule.

Example: Bayes Rules for L_p-Loss

Suppose that $L(\theta, \tilde{\theta}) = |\theta - \tilde{\theta}|^p$. If $p = 2$ (squared error loss) we need to minimize $\mathbf{E}_{\theta}(\theta - \tilde{\theta})^2$. It is easy to see that the minimum occurs at $\tilde{\theta} = \mathbf{E}(\theta)$.

Given an observation x, the minimizing rule is $\tilde{\theta}(x) = \mathbf{E}(\theta|X = x)$. In other words, the Bayes rule for squared error loss is the current mean (prior mean before data are collected, posterior mean after we see x) of θ. It is not very hard to verify that for $p = 1$ the Bayes rule is the median of the current distribution. When $p = \infty$ the loss is not finite, so we modify it to be $L(\theta, \tilde{\theta}) = 1(\theta \neq \tilde{\theta})$. The Bayes rule for this zero-one loss is the maximum of the current density. It is often called the MPD estimator (for maximum posterior density) when used on data. The MPD estimator is always contained in an HPD credibility set, whereas other Bayes rules may not be. □

In this chapter we first study several cases of parametric models, and look at inference about different parameters that may be of interest. We present general results only in regard to the one-parameter exponential family of offspring distributions, but illustrate several different techniques that can be used in a Bayesian analysis of branching process data. We then continue to look at the nonparametric case. We shall see that nonparametric is a misnomer: these models have infinite-dimensional parameters. Nonetheless some results can be obtained.

2. PARAMETRIC MODELS

We will start by considering the power series family of offspring distributions. Let $f(x|\theta) = a_x\theta^x/A(\theta)$, where $A(\theta) = \Sigma_x a_x\theta^x$. This is an exponential family with natural parameter $\log\theta$ and $M(\theta) = \log A(\theta)$. Recall that $m(\theta) = \theta A'(\theta)/A(\theta)$. A conjugate prior is given by

$$f(\theta) = \frac{\theta^\alpha}{C(\alpha, \beta)A(\theta)^\beta},$$

where $C(\alpha, \beta) = \int \theta^\alpha A(\theta)^{-\beta}\, d\theta$. The likelihood (i.e., the model distribution) based on observing $z_0, z_1, \ldots, z_n$ is proportional to $\theta^{y_n - z_0}A(\theta)^{-y_{n-1}}$. Therefore the posterior is proportional to

$$\frac{\theta^{\alpha + y_n - z_0}}{C(\alpha, \beta)A(\theta)^{\beta + y_{n-1}}}.$$

By the conjugate form, we see that the posterior density must be

$$f(\theta|\mathcal{F}_n) = \frac{\theta^{\alpha + y_n - z_0}}{C(\alpha + y_n - z_0, \beta + y_{n-1})A(\theta)^{\beta + y_{n-1}}}. \tag{1}$$

Proposition 4.2: Writing $\alpha' = \alpha + y_n - z_0$ and $\beta' = \beta + y_{n-1}$ the posterior (1) is

$$\mathbf{E}(m(\theta)|\mathcal{F}_n) = \sum_k k a_k \frac{C(\alpha' + k, \beta' + 1)}{C(\alpha', \beta')}.$$

Proof: Compute

$$\mathbf{E}(m(\theta)|\mathcal{F}_n) = \int \frac{\theta A'(\theta)}{A(\theta)} \frac{\theta^{\alpha'}}{C(\alpha', \beta')A(\theta)^{\beta'}} \, d\theta$$

and use the definition of C. □

Example: The Poisson Distribution

In the Poisson case, where $A(\theta) = \exp(\theta)$, we see that $C(\alpha, \beta) = \Gamma(\alpha + 1) \times \beta^{-(\alpha+1)}$, and that

$$\frac{C(\alpha + k, \beta + 1)}{C(\alpha, \beta)} = \left(\frac{\beta}{\beta + 1}\right)^{\alpha+1} \left(\frac{1}{\beta + 1}\right)^k \frac{\Gamma(\alpha + k + 1)}{\Gamma(\alpha + 1)}$$

so that

$$\sum ka_k \frac{C(\alpha + k, \beta + 1)}{C(\alpha, \beta)} = \left(\frac{\beta}{\beta + 1}\right)^{\alpha+1} \sum k \frac{\Gamma(\alpha + k + 1)}{\Gamma(\alpha + 1)\Gamma(k + 1)} \left(\frac{1}{\beta + 1}\right)^k .$$

This is the expected value of a negative binomial random variable, and therefore sums to $(\alpha + 1)/\beta$. A posteriori we therefore get

$$\mathbf{E}(m(\theta)|\mathcal{F}_n) = \frac{\alpha + y_n - z_0 + 1}{\beta + y_{n-1}} ,$$

which also could have been obtained directly from the fact that the posterior distribution is a gamma distribution. □

We can easily determine the MPD estimator $\tilde{m}$ of m by differentiating the logarithm of the posterior density (1). It is given by

$$\tilde{m} = \frac{\alpha + y_n - z_0}{\beta + y_{n-1}} .$$

Notice that $\tilde{m}$ does not depend on which power series family of offspring distributions we are using. As α and β approach 0, $\tilde{m} \to \hat{m}$. In order for $\tilde{m} > 1$ on a path that becomes extinct, we must have that $\alpha + y - z_0 > \beta + y$, where y is the final family tree size, i.e., $\alpha > \beta + z_0$. The amount by which $\tilde{m}$ exceeds 1 will depend on the size of y. But it is the strength of the prior that determines whether or not we will estimate a supercritical offspring distribution based on an extinct path.

Example: The Geometric Distribution

For a geometric distribution we have that $A(\theta) = (1 - \theta)^{-1}$. The conjugate prior is beta. The condition $\alpha > \beta + z_0$ means that the prior must be skewed to the right, i.e., favor large values of θ, or large values of $m(\theta) = \theta/(1 - \theta)$.

To see this more clearly, we can compute

$$\mathbf{P}\{m>1|\mathcal{F}_n\} = \int_{\theta/(1-\theta)>1} \frac{\theta^{\alpha+y_n-z_0}(1-\theta)^{\beta+y_{n+1}}}{B(\alpha+1+y_n-z_0,\ \beta+1+y_{n-1})}\, d\theta$$

$$= \int_{1/2}^{1} \frac{\theta^{\alpha+y_n-z_0}(1-\theta)^{\beta+y_{n-1}}}{B(\alpha+1+y_n-z_0,\ \beta+1+y_{n-1})}\, d\theta\ .$$

As an illustration, a geometric branching process with $\theta = 0.5$ and $z_0 = 1$ was simulated, resulting in generation sizes 1, 1, 2, 1, and 0, so that $y = 6$. Using a relatively sharp prior with $\alpha = 5$, and $\beta = 3$, so the prior mean is 1.67, we obtain $\tilde{m} = 1.11$ and

$$\mathbf{P}(m>1|\mathcal{F}_n) = \int_{1/2}^{1} \frac{\theta^{10}(1-\theta)^9}{B(11,10)}\, d\theta = 0.59\ .$$

We do not have to take into account in any special fashion that the path has become extinct. □

Application: Smallpox Epidemics in Europe

We return to the European smallpox data discussed in section 2.12. The ambiguity in the conditional approach described there is easily resolved in the present context: a Bayesian would always condition on all the data, and does not have to worry about which relevant subset of the parameter space we are in. Briefly, the conjugate prior proportional to $\lambda^{\alpha}\exp(-\beta\lambda)$ yields the posterior $\pi(\lambda|y) \propto \lambda^{\alpha+y-z_0}\exp(-\lambda[y+\beta])$. Using the interpretation of parameters suggested in the Corollary to Proposition 4.1 we can think of β as describing the total expected number of cases, with the opinion based on the equivalent of α ancestors. Basing our prior on data from 1951–1960, where there are 21 importations with a total 371 cases, for a mean of $350/371 = 0.94$, we found it reasonable to assign a prior mean of 0.94, with a strength of belief corresponding to about 50 ancestors (taking into account the possibility that the two decades are likely to be somewhat dissimilar due to improved public health policies), i.e., $\beta = 50$, $\alpha = 46$ so that

$$\mathbf{E}(\lambda|y=391,\ z_0=28) = \frac{\alpha+y-z_0+1}{y+\beta} = 0.932\ ,$$

and a 95% posterior probability interval is about (0.84, 1.02). This is similar to the maximum likelihood analysis earlier. The posterior probability that $m>1$ is about 0.07. Figure 4.1 shows the posterior density. □

In special cases it is possible to derive the posterior distribution of other parameters of the process in explicit form. We illustrate this with an example on computing the posterior for the extinction probability q.

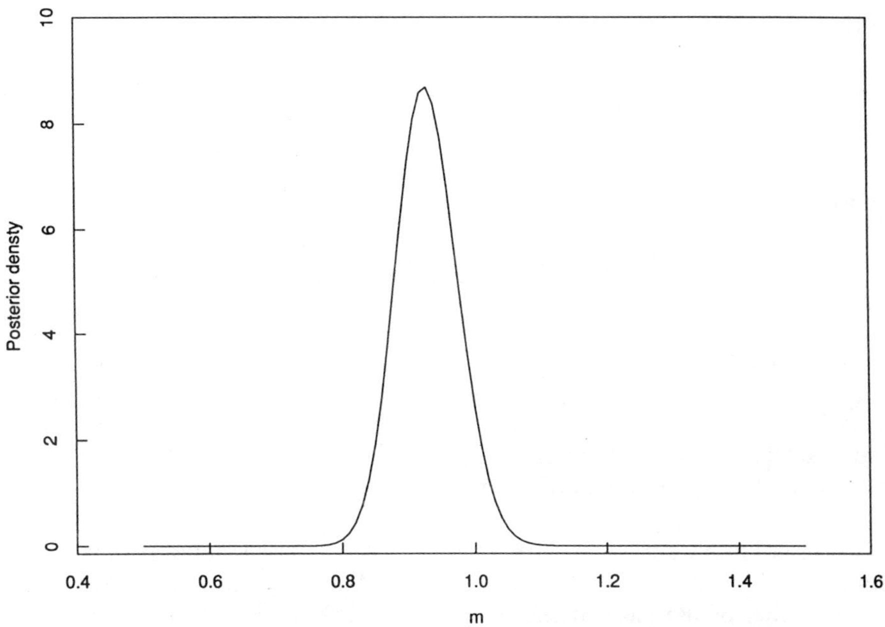

FIGURE 4.1. The posterior density of m for the European smallpox data.

Example: Extinction Probability for the Poisson Distribution

Let the offspring distribution be Poisson, and assume an exponential prior with mean $1/\theta$. Starting from $z_0 = 1$, the posterior is proportional to $\exp(-m(\theta + Y_{n-1}))m^{Y_n - 1}$, i.e., given $z_0, z_1, \ldots, z_n$ and θ, m has a gamma distribution with parameters Y_n and $\theta + Y_{n-1}$. The extinction probability $q = q(m)$ has a discrete mass of size $\mathbf{P}(m \leq 1)$ at 1. The posterior density of the continuous part of q, the solution to $\exp(-m(1-q)) = q$ for $m > 1$, is

$$f_q(x|\mathcal{F}_n) = \frac{f_m(q^{-1}(x)|\mathcal{F}_n)}{|q'(q^{-1}(x))|} .$$

Now, $q^{-1}(x)$ is obtained by solving for x the equation

$$e^{-x(1-q(x))} = q(x) \tag{2}$$

so $q^{-1}(x) = \log(x)/(x-1)$. Implicit differentiation of (2) after taking logarithms yields

$$\frac{q'(x)}{q(x)} - q(x) - xq'(x) = -1 .$$

Replacing x by $q^{-1}(x)$ we get

$$q'(q^{-1}(x)) = \frac{x-1}{\frac{1}{x} - q^{-1}(x)} = \frac{x(x-1)^2}{x-1-x\log x}\,.$$

Consequently

$$f_q(x|\mathcal{F}_n) = \frac{(\theta + Y_{n-1})^{Y_n}}{\Gamma(Y_n)}\,\frac{x\log x + 1 - x}{x(x-1)^2}\left(\frac{\log x}{x-1}\right)^{Y_n - 1} x^{-(\theta + Y_{n-1})/(x-1)}\,.$$

In Figure 4.2 we depict the density of the continuous part of the distribution of q from a path with $Y_{n-1} = 4$ and $Y_n = 6$. The prior and posterior means of m are both 1.25, corresponding to $\theta = 0.8$. The posterior mean of q is 0.67. The posterior density is bimodal, with a spike of mass 0.35 at 1 and a mode at 0.25. A 95% HPD credibility set $(0.11, 0.92) \cup \{1\}$. □

There are many situations where a parametric model is not of exponential family form, and it may be difficult to find a conjugate class. We will illustrate this in the case of the geometric distribution with modified zero term as applied to the estimation of n.

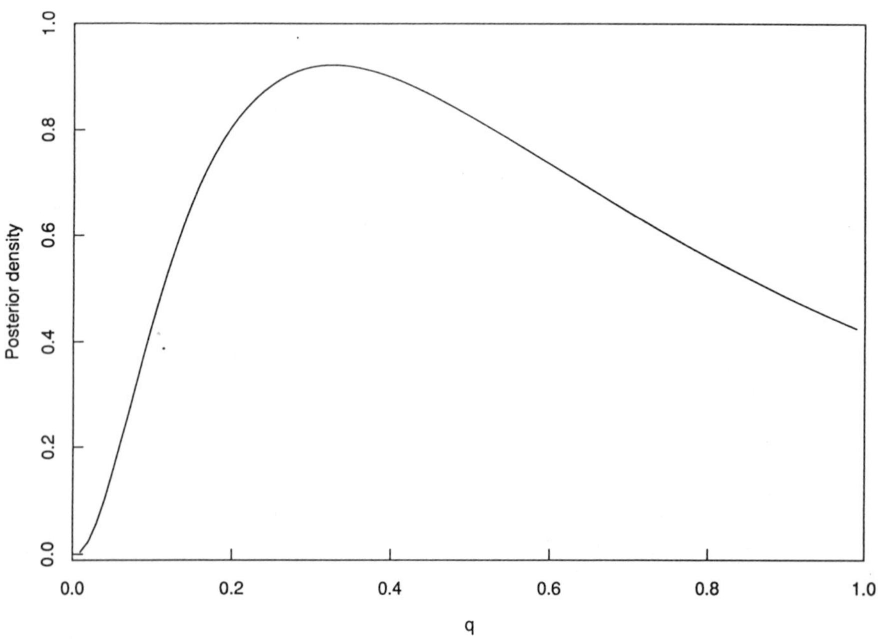

FIGURE 4.2. The continuous part of the posterior density of the extinction probability q from a simulated Poisson offspring path.

Application: Age of a Mutation

We return to the problem of estimating the age of a branching process with modified geometric offspring distribution (cf. section 3.5). The distribution of Z_n is also modified geometric with

$$\mathbf{P}(Z_n = k|n, s_0, m) = b_n c_n^{k-1}, \qquad k \geq 1,$$

where

$$b_n = m^n \left(\frac{1-s_0}{m^n - s_0}\right)^2, \qquad c_n = \left(\frac{m^n - 1}{m^n - s_0}\right),$$

$s_0 = (1-b-c)/c(1-c)$, and $m = b/(1-c)^2$. In order to investigate this problem we will assume that n is independent of the nuisance parameters m and s_0 a priori. In order to put down a reasonable prior for the nuisance parameters (given the value of n) we look at different parametrization. Instead of the parameter (s_0, m) we use (p_0, c) where $p_0 = 1 - b/(1-c)$. Since p_0 and c both correspond to probabilities, a flexible prior assumes that the two parameters have independent beta distributions. Translating back to the (s_0, m)-parametrization this corresponds to

$$f(s_0, m) \propto \frac{s_0^{\alpha-1}|1-s_0|^{\beta+\delta-1} m^{\beta-1}|1-m|^{\alpha+\gamma-1}}{|s_0 - m|^{\alpha+\beta+\gamma+\delta-1}},$$

where (α, β) are the prior parameters for p_0, and (γ, δ) for c. The region of positive density is $\{s_0 < 1, m > 1\} \cup \{s_0 > 1, m < 1\}$. We are interested in the posterior distribution of n given the observation Z_n. Since we are not directly interested in the nuisance parameters, we integrate them out, yielding

$$f(Z_n|n) = \int\int f(Z_n|n, s_0, m) f(s_0, m)\, ds_0\, dm$$

and the posterior density for n becomes

$$f(n|Z_n) = \frac{f(Z_n|n)f(n)}{\sum_n f(Z_n|n)f(n)}.$$

Turning to the Amazon Indian data, we do not have a very strong opinion about the prior distribution of n, and may be inclined to just take it flat. This is of course not a proper prior distribution, but it allows us to compute a MPD estimate of n by just maximizing $f(Z_n|n)$. Assigning a nuisance parameter prior with $(\alpha, \beta) = (15, 8)$ and $(\gamma, \delta) = (14, 7)$ yields a posterior distribution for n as depicted in Figure 4.3. The MPD estimate of n is 162, corresponding to about 5000 years, and a 95% HPD credibility interval is

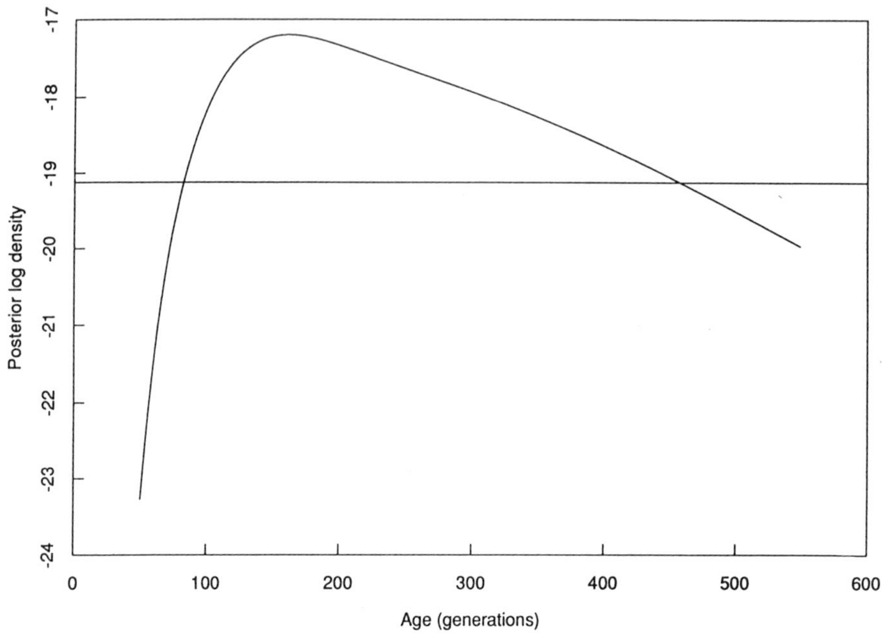

FIGURE 4.3. Posterior density for the Yanomana mutation age.

(83, 457). The horizontal line in the figure corresponds to the level of 95% of density. The interval is wider than the corresponding likelihood interval, illustrating our uncertainty about the parameters m and c. □

3. APROXIMATING THE POSTERIOR DENSITY

In the iid situation Bayes methods and likelihood methods are asymptotically equivalent (under mild regularity conditions). This is because the prior gets overwhelmed by the likelihood as the number of observations grow. Since the log likelihood (under the usual regularity conditions) behaves like a quadratic function near its maximum, the likelihood should be proportional to a normal density. Therefore one also expects the posterior distribution to be approximately normal in large samples on the explosion set. In this section we will look at two approximation results: one asserting that the posterior density is asymptotically normal on the explosion set, and a method of saddlepoint type to approximate the posterior density beyond the normal approximation.

There are quite general results regarding the normal approximation of posterior densities, often called Bernstein–von Mises theorems. The following theorem applies to any discrete-valued discrete time stochastic process satisfying the regularity conditions. We first set down some notation. Let

$L_n(\theta)$ be the likelihood, $l_n(\theta) = \log L_n(\theta)$ the twice differentiable log likelihood, and $j_n = (-J_n(\hat{\theta}))^{-1/2}$ where $\hat{\theta}$ maximizes the likelihood and J_n as before is the second derivative of the log likelihood. Finally, let $f(\theta)$ be a prior density. Here are the regularity conditions:

(i) $\int f(\theta)L_n(\theta) < \infty$ for all n

(ii) for any $\delta > 0$

$$\limsup_{n\to\infty} j_n^2 \sup_{s:\,|s-\theta|\geq\delta} (l_n(s) - l_n(\theta)) < 0$$

almost surely on $\{j_n \to 0\}$.

(iii) for each $\epsilon > 0$ there is a $\delta > 0$ such that

$$\limsup_{n\to\infty} j_n^2 \sup_{s:\,|s-\theta|<\delta} |l_n''(s) - l_n''(\theta)| < \epsilon$$

almost surely on $\{j_n \to 0\}$.

Remark: Conditions (i) and (ii) guarantee the convergence of $\hat{\theta}$ to some value θ_0 almost surely on $\{j_n \to 0\}$. □

Proposition 4.3: If $f(\theta)$ is continuous with $f(\theta_0) \neq 0$, then assumptions (i), (ii), and (iii) imply that

$$\frac{f(\hat{\theta} + j_n t)L_n(\hat{\theta} + j_n t)}{j_n \int f(s)L_n(s)\,ds} \to \frac{\exp(-t^2/2)}{(2\pi)^{1/2}}$$

almost surely on $\{j_n \to 0\}$.

Proof: All statements are made on $\{j_n \to 0\}$. A Taylor expansion shows that

$$l_n(\hat{\theta} + j_n t) - l_n(\hat{\theta}) = \tfrac{1}{2} j_n^2 t^2 l_n''(\theta^*)\,, \tag{3}$$

where $|\theta^* - \hat{\theta}| \leq j_n t$. By (iii) $j_n^2 l_n''(\theta^*) \to 1$. Hence

$$\frac{L_n(\hat{\theta} + j_n t)}{L_n(\hat{\theta})} \to \exp(-t^2/2)$$

uniformly for bounded t. By continuity $f(\hat{\theta} + j_n t)/f(\hat{\theta}) \to 1$, also uniformly for bounded t. We must show that

$$\frac{j_n \int f(s)L_n(s)\,ds}{f(\hat{\theta})L_n(\hat{\theta})} \to (2\pi)^{1/2}\,.$$

First write

$$\frac{j_n \int w(s)L_n(s)\,ds}{f(\hat{\theta})L_n(\hat{\theta})} = \int \frac{f(\hat{\theta}+j_n t)L_n(\hat{\theta}+j_n t)}{f(\hat{\theta})L_n(\hat{\theta})}\,dt\,. \tag{4}$$

For bounded sets E, the integral on the right-hand side of (4), taken over E, converges by dominated convergence to $\int_E \exp(-t^2/2)\,dt$. Split E^c into $F_{\delta,n} = \{t: |\hat{\theta}+j_n t-\theta| < \delta\} \cap E^c$ and $F_n = E^c \backslash F_{\delta,n}$. We must show that the integrals over $F_{\delta,n}$ and F_n converge to zero as $E \uparrow \mathbf{R}$ and $n \to \infty$. For the integral over $F_{\delta,n}$ it suffices to verify that

$$\lim_{E\uparrow \mathbf{R}} \lim_{n\to\infty} \int_{F_{\delta,n}} \frac{L_n(\hat{\theta}+j_n t)}{L_n(\hat{\theta})}\,dt \to 0\,.$$

But this follows from the expansion (3) and assumption (iii), which bound the integrand by $\exp(-ct^2)$ for some $c > 0$. For the final integral we use the estimate

$$\lim_{n\to\infty} \int_{t:\,|\hat{\theta}+j_n t-\theta|\ge\delta} \frac{w(\hat{\theta}+j_n t)}{w(\hat{\theta})}\,\frac{L_n(\hat{\theta}+j_n t)}{L_n(\hat{\theta})}\,dt \to 0\,.$$

Write

$$\frac{L_n(\hat{\theta}+j_n t)}{L_n(\hat{\theta})} = \frac{L_1(\hat{\theta}+j_n t)}{L_1(\hat{\theta})}\,\frac{L_{n-1}^{x_1}(\hat{\theta}+j_n t)}{L_{n-1}^{x_1}(\hat{\theta})}$$

where $L_{n-1}^{x_1}$ is the conditional likelihood, given the first observation. Let $l_{n-1}^x = \log L_{n-1}^x$. For $\eta > 0$ we have

$$\begin{aligned}
&\limsup_{n\to\infty} j_n^2 \sup_{s:\,|s-\theta|>\eta} (l_{n-1}^{x_1}(s) - l_{n-1}^{x_1}(\theta)) \\
&= \limsup_{n\to\infty} j_n^2 \sup_{s:\,|s-\theta|>\eta} (l_n(s) - l_n(\theta) + l_1(\theta) - l_1(s) \\
&\le \limsup_{n\to\infty} j_n^2 \sup_{s:\,|s-\theta|>\eta} (l_n(s) - l_n(\theta)) + \limsup_{n\to\infty} j_n^2 \sup_{s:\,|s-\theta|>\eta} (l_1(\theta) - l_1(s)) \\
&= \limsup_{n\to\infty} j_n^2 \sup_{s:\,|s-\theta|>\eta} (l_n(s) - l_n(\theta)) < \theta\,.
\end{aligned}$$

Hence for any t such that $|\hat{\theta}+j_n t-\theta| \ge \delta$

$$L_{n-1}^{x_1}(\hat{\theta}+j_n t)/L_{n-1}^{x_1}(\hat{\theta}) \le \exp(-dj_n^{-2})$$

for some $d > 0$. Consequently

$$\int_{t:\,|\hat{\theta}+j_n t-\theta|\ge\delta} \frac{f(\hat{\theta}+j_n t)}{f(\hat{\theta})}\,\frac{L_n(\hat{\theta}+j_n t)}{L_n(\hat{\theta})} \le \frac{\exp(-dj_n^{-2})}{f(\hat{\theta})L_n(\hat{\theta})j_n} \int f(s)L_1(s)\,ds \to 0\,.$$

□

Corollary 1: For a BGW process with power series offspring distribution, the posterior distribution of θ is asymptotically normal on the explosion set with asymptotic mean θ_0 and asymptotic variance $\theta_0^2/Y_{n-1}\sigma^2(\theta_0)$.

Proof: We have that

$$l_n'(\theta) = (Y_n - z_0 - Y_{n-1}m(\theta))/\theta$$

and

$$l_n''(\theta) = -(Y_{n-1}\sigma^2(\theta) + Y_n - z_0 - Y_{n-1}m(\theta))/\theta^2 ,$$

so that $j_n = Y_{n-1}m'(\hat{\theta})/\hat{\theta}$. Since m and m' are continuous, we see that (iii) is satisfied. Furthermore, since l_n has a strict maximum at $\hat{\theta}$, (ii) follows immediately. Condition (i) is trivial in this case. □

It is straightforward to turn this result into one about the offspring mean.

Corollary 2: Under the conditions of Proposition 4.3 and Corollary 1, the posterior mean is asymptotically normal with asymptotic mean $m(\theta_0)$ and asymptotic variance $\sigma^2(\theta_0)/Y_{n-1}$.

Proof: Immediate from a standard Taylor series expansion and Corollary 1. □

We see that from a frequency point of view, the asymptotic behaviors of the mle and a Bayes estimator (for any reasonable prior) are the same when we use the appropriate norming, namely the observed information.

Instead of a normal approximation it is sometimes possible to get a more accurate approximation of a posterior density or mean by approximating the integrals involved. The general idea, called the Laplace method, was outlined in a Bayesian context by Tierney and Kadane (1986). If we want to compute an integral of the form $\int \exp(nh(\theta))\,d\theta$, where h has a unique maximum at $\hat{\theta}$, then for large n the integral depends only on the behavior of h near this maximum. Replacing $h(\theta)$ by $h(\hat{\theta}) - \frac{1}{2}(\theta - \hat{\theta})^2 h''(\hat{\theta})$ we get

$$\int e^{nh(\theta)}\,d\theta \approx \sigma\left(\frac{2\pi}{n}\right)^{1/2} e^{nh(\hat{\theta})} ,$$

where $\sigma^2 = -1/h''(\hat{\theta})$. The error in this approximation can be obtained by further Taylor expansions, and is seen to be $O(n^{-1})$ (Exercise 3).

In order to apply this to a branching process, we look first at the power series case. Here the posterior density, using the conjugate prior $f(\theta)$, is

$$f(\theta|\mathcal{F}_n) = \frac{f(\theta)A(\theta)^{-Y_{n-1}}\theta^{Y_n-1}}{\int f(\theta)A(\theta)^{-Y_{n-1}}\theta^{Y_n-1}\,d\theta} = \frac{A(\theta)^{-(\beta+Y_{n-1})}\theta^{\alpha+Y_n-1}}{\int A(\theta)^{-(\beta+Y_{n-1})}\theta^{\alpha+Y_n-1}\,d\theta}.$$

We use Laplace's method to approximate the denominator, now with Y_{n-1} playing the role of n. First write

$$\int A(\theta)^{-(\beta+Y_{n-1})}\theta^{\alpha+Y_n-1}\,d\theta = \int \exp\left(Y_{n-1}\left(\frac{\alpha+Y_n-1}{Y_{n-1}}\log\theta - \left(1+\frac{\beta}{Y_{n-1}}\right)\log A(\theta)\right)\right)d\theta\,.$$

For ease of writing, set $\alpha^* = (\alpha + Y_n - 1)/Y_{n-1}$ and $\beta^* = 1 + \beta/Y_{n-1}$. We have

$$h(\theta) = \alpha^* \log\theta - \beta^* \log A(\theta)\,,$$

which is maximized at the solution θ^* to

$$\theta A'(\theta)/A(\theta) = \alpha^*/\beta^*\,.$$

Thus we approximate the integral by

$$c^* = \left(\frac{2\pi}{Y_{n-1}}\right)^{1/2} \sigma^* A(\theta^*)^{-\beta^* Y_{n-1}}\theta^{*\alpha^* Y_{n-1}}\,,$$

where $(\sigma^*)^{-2} = \alpha^*/\theta^{*2} + \beta^* D^2 \log A(\theta^*)$, and approximate the posterior density by

$$f(\theta|\mathcal{F}_n) \approx (c^*)^{-1}\theta^{\alpha^* Y_{n-1}}A(\theta)^{-\beta^* Y_{n-1}}\,.$$

Example: Poisson Offspring Distribution

Here $\theta^* = \alpha^*/\beta^*$ and $\sigma^{*2} = \beta^*/\theta^*$, so

$$c^* = (2\pi)^{1/2}(\theta^*)^{Y_{n-1}\alpha^*+1/2}e^{-Y_{n-1}\beta^*}/(Y_{n-1}\beta^*)^{1/2}\,.$$

The exact integral is

$$c = \Gamma(Y_{n-1}\alpha^* + 1)(Y_{n-1}\beta^*)^{-Y_{n-1}\alpha^*+1}\,.$$

Using Stirling's formula $\Gamma(x+1) \sim (2\pi)^{1/2}x^{x+1/2}e^{-x}$ we see that c^* is precisely the Stirling approximation to c. □

The Laplace approximation is particularly good when computing posterior means. Here the approximation is

$$\mathbf{E}(g(\theta)|\mathcal{F}_n) \approx \frac{\sigma^*}{\sigma} \exp(-n(h^*(\hat{\theta}^*) - h(\hat{\theta}))) ,$$

where $h^*(\theta) = (\log g(\theta) + \log f(\theta) + \log L_n(\theta))/n$ and $h(\theta)$ is the same function without the $\log g(\theta)$. The quantities $\hat{\theta}^*$ and $\hat{\theta}$ maximize the respective functions. To get an idea of why this approximation works so well, first note that the posterior mean $\mathbf{E}(g(\theta)|\mathcal{F}_n)$ is a ratio of two integrals:

$$\mathbf{E}(g(\theta)|\mathcal{F}_n) = \frac{\int g(\theta) f(\theta) L_n(\theta)\, d\theta}{\int f(\theta) L_n(\theta)\, d\theta} ,$$

where $L_n(\theta)$ is the likelihood. Approximating each integral by Laplace's method would ordinarily yield an error of order $1/n$. However, the order $1/n$ terms cancel, yielding an overall approximation error of order $1/n^2$. Intuitively, if g is bounded away from zero near the posterior mode, the shape of the integrals is similar, so the approximation in both numerator and denominator will be making similar errors. Taking the ratio of the terms, a portion of these errors will cancel.

Example: The Mean of a Geometric Offspring Distribution

Using the conjugate beta prior, we want to estimate $\theta/(1-\theta)$. Let $\alpha_n = Y_{n-1}\alpha^*$ and $\beta_n = Y_{n-1}\beta^*$. Then

$$Y_{n-1}h(\theta) = \alpha_n \log\theta + \beta_n \log(1-\theta)$$

and

$$Y_{n-1}h^*(\theta) = (\alpha_n + 1)\log\theta + (\beta_n - 1)\log(1-\theta) ,$$

so that $\hat{\theta} = \alpha_n/(\alpha_n + \beta_n)$, and $\hat{\theta}^* = (\alpha_n + 1)/(\alpha_n + \beta_n)$. Furthermore,

$$\sigma^{*2} = \frac{Y_{n-1}}{\dfrac{\alpha_n + 1}{\hat{\theta}^{*2}} + \dfrac{\beta_n - 1}{(1-\hat{\theta}^*)^2}}$$

and

$$\sigma^2 = \frac{Y_{n-1}}{\dfrac{\alpha_n}{\hat{\theta}^2} + \dfrac{\beta_n}{(1-\hat{\theta})^2}} .$$

We therefore get the approximate posterior mean

$$\mathbf{E}(m|\mathcal{F}_n) \approx \frac{\sigma^*}{\sigma}\left(1+\frac{1}{\alpha_n}\right)^{\alpha_n}\left(1-\frac{1}{\beta_n}\right)^{\beta_n}\frac{\hat{\theta}^*}{1-\hat{\theta}^*}\,.$$

For large values of Y_{n-1} a Taylor expansion yields

$$\mathbf{E}(m|\mathcal{F}_n) = \frac{\alpha_n+1}{\beta_n-1} + O_{\mathbf{P}}(Y_{n-1}^{-1})\,,$$

which is within $O_{\mathbf{P}}(Y_{n-1}^{-1})$ of the correct answer α_n/β_n. For a numerical example, a path with mean 1.5 was simulated, with $Y_{n-1}=53$ and $Y_n=77$. Using the prior with $\alpha=2$ and $\beta=1$, corresponding to a prior mean of 2, the Laplace approximation yields $\mathbf{E}(m|\mathcal{F}_n)\approx 1.47$, where the true posterior mean is 1.44. □

Application: Australian Influenza Importations

During the great influenza epidemic at the end of World War I, the Australian government kept careful track of the epidemics occurring in incoming ships during their voyage to Australia. In a much overlooked paper, McKendrick (1926), the inventor of the birth and death process and several other standard models of applied probability, discussed these data, although he was not able to analyze them using his methods.

We will assume that all the epidemics had run their course (and become extinct) at the arrival, and that a branching process model with Poisson offspring distribution applies. The data are then the total sizes of these processes:

Number of cases	1	2	3	4	5	6	7	8	9	>9
Number of ships	34	15	5	3	8	1	3	4	2	17

As seen in section 2.12, the distribution of total size for a Poisson offspring distribution is the Borel–Tanner distribution. The fit of the data above to a Borel–Tanner distribution is quite good, with a chi-square test P-value of 0.285.

Since $f(x|\theta)\propto\theta^{x-1}\exp(-\theta x)$, a conjugate prior is a gamma prior. Notice however that the data have been grouped, so that the likelihood contains some terms which are tail probabilities of the distribution. We will use a prior with mean 1 and strength five observations, i.e., $\alpha=\beta=5$. Since an exact computation requires numerical integration, we want an approximation of the posterior mean. The Laplace approximation only involves maximizing two functions, namely

$$nh^*(\theta) = (T - N + \alpha)\log\theta - (T + \beta)m + (n - N)\log(1 - F(9|\theta))$$

and

$$nh(\theta) = (T - N + \alpha - 1)\log\theta - (T + \beta)m + (n - N)\log(1 - F(9|\theta))\,,$$

where $F(x|\theta)$ is the cdf of the Borel–Tanner distribution and T is the sum of the N values that are known precisely. We estimate σ^* and σ by numerical differentiation of the respective functions, yielding $\hat{\theta} = 0.9054$, $\hat{\theta}^* = 0.9088$, $\sigma = 0.0558$, and $\sigma^* = 0.0560$, and the posterior mean 0.9092. In terms of sizes of epidemics, this corresponds to an average size of 11.0, with a standard deviation of 34.8. The Borel–Tanner distribution is very long-tailed. □

4. A NONPARAMETRIC MODEL

We now turn to the nonparametric case, assuming only that the offspring distribution is $(p_k, k \geq 0)$. If the support of the distribution is finite, we are essentially in a multinomial situation (although the sample size is random). Therefore we will start by looking at Bayes estimation of a multinomial parameter from a fixed size sample. Let $\xi_1, \ldots, \xi_n$ be observations from a multinomial distribution on $K+1$ categories with probability vector $p_0, \ldots, p_K$. The likelihood is proportional to $\Pi_{0 \leq i \leq K}\, p_i^{n_i}$ where

$$n_i = \#\{\xi_j = i,\ j = 1, \ldots, n\}\,.$$

We see that a conjugate family of prior densities is of the form

$$f(p_0, \ldots, p_K) = c(\alpha)\prod_{i=0}^{K} p_i^{\alpha_i - 1}\,, \qquad p_i \geq 0,\ \sum p_i = 1,\ \alpha_i > 0\,.$$

If $\alpha_i = 0$, the corresponding $p_i = 0$. It is an exercise in integration to show that

$$c(\boldsymbol{\alpha}) = \frac{\Gamma(\alpha_0 + \cdots + \alpha_K)}{\Gamma(\alpha_0)\cdots\Gamma(\alpha_K)}\,.$$

This distribution is called the Dirichlet distribution with parameter $\boldsymbol{\alpha}$, and denoted $\mathrm{Dir}(\boldsymbol{\alpha})$. In the case $K = 1$ it reduces to the beta distribution, and in general it arises from a vector of independent gamma variates by rescaling each by the sum of all (Exercise 4). From this representation it is clear that if we add together (group) some p_i, we get a Dirichlet distribution on fewer categories, with the α_i added in the same fashion as the p_i.

Lemma 4.1: If $p_0, \ldots, p_K \sim \text{Dir}(\alpha_0, \ldots, \alpha_K)$, then with $\alpha = \Sigma_0^K \alpha_i$

$$\mathbf{E}p_i = \frac{\alpha_i}{\alpha},$$

$$\mathbf{Var}\ p_i = \frac{\alpha_i(\alpha - \alpha_i)}{\alpha^2(\alpha + 1)},$$

and for $i \neq j$,

$$\mathbf{Cov}(p_i, p_j) = -\frac{\alpha_i \alpha_j}{\alpha^2(\alpha + 1)}.$$

Finally if, given $p_0, \ldots, p_K$, ξ is an observation from the multinomial distribution with these probabilities, then

$$(p_0, \ldots, p_K | \xi = x) \sim \text{Dir}(\alpha + e(x)),$$

where $e(x)$ is a vector with elements $1(i = x)$.

The proof is left as Exercise 5. Notice that we have not really used the finiteness of K in any essential way, and all the properties continue to hold if $K = \infty$, provided $\alpha = \Sigma_0^\infty \alpha_i < \infty$. We can interpret $(\alpha_0/\alpha, \alpha_1/\alpha, \ldots)$ as the prior expected distribution, with strength of belief corresponding to about α observations.

We now return to a BGW process. Assume first that we observe the entire tree for n generations, and write as before $n_j = \Sigma_{m=1}^n \#\{X_{m,l} = j, l \leq Z_{m-1}\}$. By the lemma, denoting the σ-algebra of the entire tree up to time n by $\mathcal{F}_n^X$, we deduce that

$$\tilde{p}_i = \mathbf{E}(p_i | \mathcal{F}_n^X) = \frac{\alpha_i + n_i}{\alpha + Y_{n-1}}.$$

This is the Bayes estimator with respect to any loss function of the form

$$L(\mathbf{p}, \mathbf{q}) = \sum_{j=0}^{\infty} w_j (p_j - q_j)^2,$$

where the w_j are positive weights.

Application: Cell Physiology

Yeast cells reproduce by budding. When a daughter cell splits off, a scar is left on the mother cell. Thus the number of offspring of any cell in a yeast colony can be determined by counting the number of scars on its surface. Gani and Saunders (1976) develop a simple model for the offspring distribution by assuming that in each time increment the cell either dies (with probability r), surviving without budding (with probability q) or buds (with probability p). Then the total proportion of cells with k offspring eventually will have a geometric distribution with probability $\frac{1}{2}$ of success, provided that the colony does not become extinct. The same offspring distribution arises from an underlying birth and death process (Saunders, 1976).

Gani and Saunders (1977) report data of James A. Barnett of the University of East Anglia on the number of cells of k scars, $k = 0, 1, \ldots, 8$, obtained experimentally from various varieties of yeast. We will first look at data on *Saccharomyces cerevisiae* during its exponential growth phase. Using the offspring distribution derived by Gani and Saunders (1976), we assign a Dirichlet prior with mean vector having $\alpha_k = \alpha 2^{-(k+1)}$. Taking α, the strength of the prior, to be 200, we obtain the following table.

***Saccharomyces cerevisiae* in exponential growth**

	$k=0$	1	2	3	4	≥ 5
n_k	287	149	98	43	16	11
α_k	100	50	25	12.5	6.25	6.25
$\tilde{p}_k$	0.481	0.248	0.153	0.069	0.028	0.021
$\text{se}(\tilde{p}_k)$	0.018	0.015	0.013	0.009	0.006	0.005

Since the marginal posterior distribution of each category is beta with large parameter values, we can approximate them quite well with normal distributions. We see that the theoretical model is well supported, in that the value $p_k = 2^{-(k+1)}$ usually falls well inside a 95% individual HPD interval. The probabilities for $k = 2$ and for the tail are 2.1 standard errors from the theoretical value.

The fit of the theoretic model for *Saccharomyces cerevisiae* colonies in its stationary phase is not as good. The prior used above yields the following results.

***Saccharomyces cerevisiae* in stationary phase**

	$k=0$	1	2	3	4	≥ 5
n_k	685	267	213	88	57	32
α_k	100	50	25	12.5	6.25	6.25
$\tilde{p}_k$	0.509	0.206	0.171	0.065	0.041	0.025
$\text{se}(\tilde{p}_k)$	0.013	0.010	0.009	0.006	0.005	0.004

Here the fit of the theoretical model for $k = 1$ and 2 is quite poor. Figure 4.4 shows the contours corresponding to 0.5, 0.75, 0.9, 0.95, and 0.99 of joint posterior probability for p_0 and p_1. The posterior density is well approximated by a bivariate normal density with mean vector $(0.509, 0.206)$ and

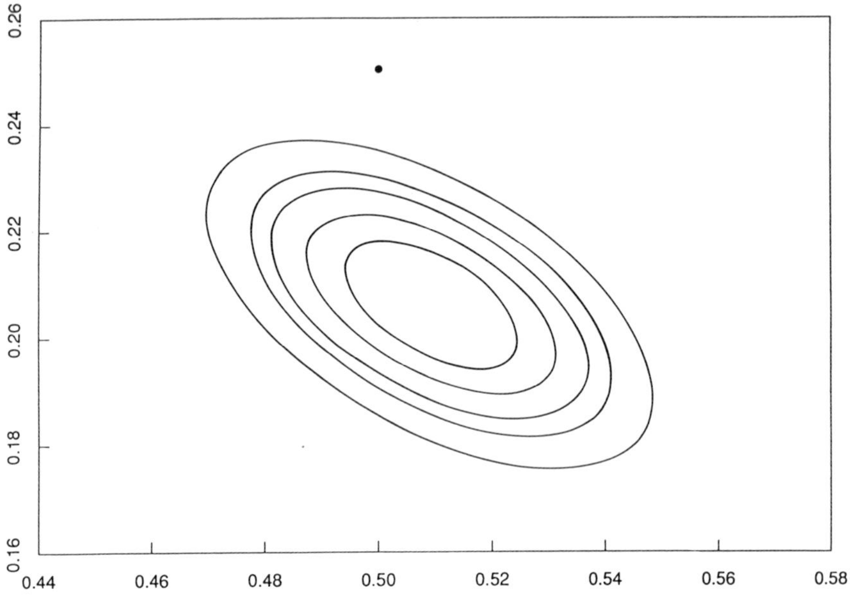

FIGURE 4.4. Posterior density contours for (p_0, p_1) of the stationary phase of yeast growth. The dot corresponds to the theoretical value (0.50, 0.25).

covariance matrix given by the standard errors in the table and correlation -0.56. The theoretical value $(0.50, 0.25)$ has very low posterior density. One explanation may be that the birth rates for newly formed cells is lower than that for mature cells. This can be modeled by assuming as a prior, a Dirichlet distribution with mean measure being a mixture of a point mass at zero (corresponding to newly formed cells) and a geometric distribution. Assuming that in the steady state about 10% of the cells are nonbudding, and using $\frac{1}{2}$ for the geometric success probability as before we get the following results.

***Saccharomyces cerevisiae* in stationary phase, mixture model**

	$k=0$	1	2	3	4	≥ 5
n_k	685	267	213	88	57	32
α_k	110	45	22.5	11.25	5.625	5.625
$\tilde{p}_k$	0.516	0.202	0.153	0.064	0.041	0.024
$\text{se}(\tilde{p}_k)$	0.013	0.010	0.009	0.006	0.005	0.004

A model with 19% nonbudding cells, and a geometric distribution with success probability 0.40 appears to fit the data adequately. However, the use of a branching process to model a cell colony in steady state is open to criticism. □

In order to get the Bayes estimate based on generation sizes only, we must compute

$$\mathbf{E}(p_i|\mathcal{F}_n^Z) = \mathbf{E}(\tilde{p}_i|\mathcal{F}_n^Z) = \frac{\alpha_i + \mathbf{E}(n_i|\mathcal{F}_n^Z)}{\alpha + Y_{n-1}} .$$

The computation of $\mathbf{E}(n_i|\mathcal{F}_n^Z)$ was discussed in section 3.3. We can estimate the offspring mean by

$$\tilde{m} = \sum i\tilde{p}_i = \frac{\alpha m_0 + Y_n - z_0}{\alpha + Y_{n-1}} = \frac{\alpha}{\alpha + Y_{n-1}} m_0 + \frac{Y_{n-1}}{\alpha + Y_{n-1}} \hat{m} ,$$

where m_0 is the expected prior mean $\Sigma\, i\alpha_i/\alpha$. This is very similar to the corresponding result for the parametric case of a power series offspring distribution in section 1.

It is sometimes important to determine the exact distribution of (a function of) the mean of a Dirichlet distribution, as opposed to just knowing the mean of the mean, m_0. The following result is due to Hannum et al. (1981). In this special case it is a trivial consequence of Exercise 4.

Lemma 4.2: Let $m = \Sigma\, ip_i$ be the mean of a distribution randomly chosen from the Dirichlet distribution with parameter $\alpha_0, \alpha_1, \ldots$. Then

$$\mathbf{P}(m \le x) = \mathbf{P}\left(\sum_{j=0}^{\infty} (j - x)\xi_j \le 0\right),$$

where $\xi_j \sim \Gamma(\alpha_j, 1)$.

5. SOURCES

The brief development of Bayesian theory in section 1 uses the work by Diaconis and Ylvisaker (1979) and is otherwise fairly standard.

The parametric Bayes theory has not been pursued much. Jagers (1975, section 2.13) treats the binary splitting case. Some standard distributions were dealt with by Dion (1972) in his unpublished Ph.D. thesis. Heyde (1979) and Scott (1987) have also done some work in this area.

Approximation methods for posterior densities have recently been the subject of much work. A Bernstein–von Mises theorem for stochastic processes was given by Heyde and Johnstone (1979), and the presentation in section 3 follows Fraser and McDunnough (1984). Scott (1987) also contains material in this area. The development of precise approximations to posterior densities using Laplace's approximation, due to Tierney and Kadane (1986) and Tierney et al. (1987), is closely related to current work on saddlepoint approximations (see Reid, 1988, and the associated discussion). The application to influenza importations in Australia is based on a paper by McKendrick (1926), arguably the most overlooked applied probabilist in the history of the subject.

The nonparametric Bayes approach is due to Johnson et al. (1979), who developed the theory in a more general context of age-dependent branching processes. Again, Dion (1972) did some work in this area in his thesis. Gani and Saunders (1976, 1977) contain the data and some models for the yeast cell application.

EXERCISES

1. Show that the family of finite mixtures of conjugate prior distributions is a conjugate family.

2. Find a class of conjugate prior distributions for the negative binomial offspring distribution, and determine the Bayes estimator of m for squared error loss.

3. Discuss the estimation of σ^2 from iid observations of a one-parameter exponential family with the natural conjugate prior distribution.

4. Show that when f is unimodal, the HPD interval minimizes the length among all $(1-\alpha)$-credibility intervals.
   Hint: Minimize $\int_{l(x)}^{u(x)} 1\,dx$ subject to $\int_{l(x)}^{u(x)} f(\theta|x)\,d\theta \geq 1-\alpha$.

5. For the data in the Application of section 2, find an HPD credibility set for q^n.

6. Applying Proposition 2 to the geometric distribution, deduce the combinatorial identity

$$\sum_{k=1}^{\infty} \frac{k(\alpha+1)\cdots(\alpha+k)}{(\alpha+\beta+2)\cdots(\alpha+\beta+k+2)} = \frac{\alpha+1}{\beta(\beta+1)}\,.$$

7. (a) Let Y be the total number of individuals in an extinct branching process with power series offspring distribution. Show that Y also has a power series offspring distribution with parameter $\phi = \theta/A(\theta)$ and normalizing constant $B(\phi) = \theta^n$, where n is the final nonzero generation.
   Hint: Write an equation for the pgf of Y_n in terms of previous generations.
   (b) For the geometric case, show that the total size has a negative binomial distribution.
   (c) Compare the fit of the epidemics data in section 3 to a Poisson offspring distribution with that to a geometric offspring distribution.
   (d) Find the posterior probability of supercriticality for the data in (c) using a geometric offspring distribution.

8. Prove that the error in the Laplace method is $O(n^{-1})$.
Hint: Use a Taylor expansion to several terms of the exponent in the integrand.

9. (a) Show that

$$\lim_{n\to\infty}(\mathbf{P}(m>1|Z_n>0)-(1-\Phi((1-\tilde{m})/(\sigma Y_{n-1}^{1/2}))))=0\,.$$

(b) Apply this result to the data in the last Example of section 3.

10. Let $Y_1,\ldots,Y_n$ be iid $\Gamma(\alpha_i,1)$. Show that

$$\left(\frac{Y_1}{\sum Y_i},\ldots,\frac{Y_n}{\sum Y_i}\right)\sim \mathrm{Dir}(\alpha_1,\ldots,\alpha_n)\,.$$

11. Let $X_1,\ldots,X_k\sim \mathrm{Dir}(\alpha_1,\ldots,\alpha_k)$.
(a) Show that

$$\mathbf{E}\{X_1^{\delta_1}\cdots X_k^{\delta_k}\}=\frac{\Gamma\left(\alpha+\sum_1^k\delta_i\right)}{\Gamma(\alpha)}\prod_{i=1}^k\frac{\Gamma(\alpha_i+\delta_i)}{\Gamma(\alpha_i)}$$

(b) Deduce Lemma 4.1.

12. The following data are offspring numbers for female ancestors of Thule eskimos carrying the gene for blood group B (Gilberg et al., 1978; Sheehan, 1990). It may be reasonable to regard this as a random sample from the eskimo pedigree, conditional upon offspring.

Offspring number	1	2	3	4	5
Number of mothers	6	4	2	3	1

Estimate the offspring distribution (in a nonparametric fashion) and sketch the posterior density of the conditional offspring mean given nonzero offspring.

CHAPTER 5

Some Variants on the BGW Branching Process

The simple BGW process is not always adequate to describe actual phenomena. There are many slight variants that have been proposed to deal with particular problems. In this chapter we discuss a few of these, and present applications for which the modification is necessary. A process with immigration is applied to the study of physical Brownian motion; varying environment is needed to describe the spread of rumors; a continuous state space model describes plasma progesterone concentration for pregnant women; a simple multiple process is used to study the migration of insects. In each case some relevant statistical theory is outlined.

1. A BRANCHING PROCESS WITH IMMIGRATION

Let $Y_1, Y_2, \ldots$ be iid random variables with pgf $Q(s) = \Sigma\, a_k s^k$, and define a *branching process with immigration* recursively by

$$Z_k = \sum_{1}^{Z_{k-1}} X_{k,i} + Y_k ,$$

where as before the $X_{k,i}$ are iid with pgf $P(s)$, independently of the Y_k. Let m and σ^2 be the mean and variance of P, and (when they exist) let κ and β^2 be those of Q. The process starts with the first nonzero Y_k. Clearly the pgf for Z_n satisfies

$$P_n(s) = \mathbf{E}s^{Z_n} = \mathbf{E}(P(s)^{Z_{n-1}})Q(s)$$

whence

$$P_n(s) = Q(s)P_{n-1}(P(s)) . \tag{1}$$

We can think of Z_n as constructed by summing up many independent branching processes, each started at the time of an immigration. If each of these processes are subcritical, so that they are a.s. finite in length, it seems reasonable that Z_n should converge to a limit, i.e., that the process would have a stationary distribution. Letting $n \to \infty$ in (1) we see that if Z_n converges to a limiting distribution, the pgf $\pi(s)$ of this distribution must satisfy

$$\pi(s) = Q(s)\pi(P(s)) .$$

In the supercritical case the immigration plays a much less important role, and Z_n behaves much more like a BGW process does on the set of nonextinction. In other words, neither in the subcritical nor the supercritical case does the process become extinct.

Let $\mathcal{F}^Y = \sigma(Y_0, Y_1, \ldots)$, $W_n = m^{-n}Z_n$, and $S = \Sigma_{n=0}^{\infty} m^{-n}Y_n$.

Proposition 5.1

(i) When $1 < m < \infty$, W_n converges a.s. on the set $\{S < \infty\}$.

(ii) The condition

$$\mathbf{E}\log_+ Y_1 = \sum_{k=1}^{\infty} a_k \log k < \infty \tag{2}$$

implies that $\mathbf{P}(S < \infty) = 1$.

(iii) When $0 < m < 1$ then Z_n converges in distribution whenever (2) holds.

(iv) When $m = 1$, $\sigma^2 < \infty$, and the Y_i are iid, then $Z_n \xrightarrow{d} \infty$.

Proof: To prove (i), first note that

$$\mathbf{E}(Z_{n+1}|Z_n, \mathcal{F}^Y) = mZ_n + Y_{n+1} \geq mZ_n ,$$

so that W_n is a nonnegative submartingale with respect to $\mathcal{F}_n \wedge \mathcal{F}^Y$, where $\mathcal{F}_n$ is $\sigma(Z_0, \ldots, Z_n)$. Hence it converges a.s. on the set $\{\sup \mathbf{E}(W_n|\mathcal{F}^Y) < \infty\}$. But $\mathbf{E}(W_n|\mathcal{F}^Y) = \Sigma_0^n m^{-k}Y_k$, so W_n converges a.s. on $\{S < \infty\}$.

In order to verify (ii), let $1 < \mu \leq m$, and compute

$$\sum_0^{\infty} \mathbf{P}(Y_n > \mu^n) = \sum_{k=1}^{\infty}\sum_{n=0}^{\infty} 1(k > \mu^n)a_k = \sum_{k=1}^{\infty}\left[\frac{\log k}{\log \mu}\right]a_k = O(\mathbf{E}\log_+ Y_1) .$$

where $[\cdot]$ stands for the integer part. Hence, if (2) holds, $\Sigma\, \mathbf{P}(Y_n > \mu^n) < \infty$, so by Borel–Cantelli we have $Y_n \leq \mu^n$ eventually, a.s. Choosing $\mu < m$, we see that the tail of the series $\Sigma\, m^{-n}Y_n$ eventually will be dominated by the tail of $\Sigma\, (\mu/m)^n < \infty$.

To prove (iii), write

$$Z_n = \sum_{k=0}^{n} \sum_{l=1}^{Y_k} Z_{n-k}^{k,l},$$

where $(Z_{\cdot}^{k,l})$ are iid BGW processes starting at time k. Since their distributions do not depend on k, l, and since $(Y_0, \ldots, Y_n) \stackrel{d}{=} (Y_n, \ldots, Y_0)$, we have that

$$Z_n \stackrel{d}{=} \hat{Z}_n = \sum_{k=0}^{n} \sum_{l=1}^{Y_k} Z_k^{k,l}.$$

Hence Z_n converges in distribution iff $\hat{Z}_n$ does, i.e., since $\hat{Z}_n$ is increasing, iff $\hat{Z}_\infty < \infty$ a.s. In order for this to happen we can have $Z_k^{k,l} > 0$ for only finitely many k, l with $l \leq Y_k$. Conditioning upon $\mathcal{F}^Y$ and using Borel–Cantelli, this will hold iff $\Sigma\,(1 - q_n)Y_n < \infty$, where q_n is the extinction probability at time n of any of the $(Z_{\cdot}^{k,l})$. Using Exercise 2.9(b) we know that $1 - q_n = O(m^n)$, whence, from (ii), we see that $\Sigma\, q_n Y_n < \infty$ whenever (2) holds.

Finally, (iv) follows from the Corollary to Proposition 2.9, stating that $1 - q_n = O(1/n)$ if $m = 1$ and $\sigma^2 < \infty$. But $\Sigma\, Y_n/n = \infty$ with probability one when the Y_i are iid, and the argument in (iii) shows that Z_n therefore does not converge in distribution. □

Remark: If $m = 1$ and $\sigma^2 < \infty$, then the resulting Markov chain (Z_n) is null recurrent if $2\kappa \leq \sigma^2$, and transient if $2\kappa > \sigma^2$. □

Example: Birth and Death Process with Immigration

Let $X(t)$ be a linear birth and death process with immigration occurring according to a Poisson process. The probability of a population increase by one during $(t, t + dt)$, if $X(t) = i$, is $(i\lambda + \nu)\,dt$, and the probability of decrease by one is $i\mu\,dt$. The probability generating function $\Pi_t(s)$ for $X(t)$ is obtained by conditioning on the immigration times, which have the distribution of uniform order statistics from $(0, t)$ and starting a birth-and-death process at each immigration time (Bartlett, 1966, equation 3.41(2)):

$$\Pi_t(s) = \Phi(s; t)^{X(0)} \exp\left(-\nu \int_0^t (1 - \Phi(s; u)\,du \right),$$

where $\Phi(s; t)$ is the pgf of a linear birth-and-death process, given by

$$\Phi(s; t) = \frac{\mu\tau(s-1) + \mu - \lambda s}{\lambda\tau(s-1) + \mu - \lambda s}$$

and where $\tau = \exp((\lambda - \mu)t)$. Performing the integration we get

$$\Pi_t(s) = \left(\frac{\mu(\tau - 1) - (\mu\tau - \lambda)s}{\lambda\tau - \mu - \lambda(\tau - 1)s} \right)^{X(0)} \left(\frac{\lambda - \mu}{\lambda\tau - \mu - \lambda(\tau - 1)s} \right)^{\nu/\lambda}. \quad (3)$$

If we just look at this process at integer times, the discrete skeleton $Z_n = X(n)$ has a pgf $\pi_n = \Pi_n$. We see that π_n satisfies

$$\pi_{n+1}(s) = \exp\left(-\nu \int_0^1 (1 - \Phi(s; u))\, du\right) \pi_n(\Phi(s; 1))\,.$$

Thus the discrete skeleton has a pgf satisfying (1), so it is a branching process with immigration, where the offspring distribution is the same as for the discrete skeleton of a linear birth and death process (see the first Example in section 1.3), and the immigration distribution has pgf $Q(s) = \exp(-\nu \int_0^1 (1 - \Phi(s; u))\, du)$, with mean $\nu(\exp(\lambda - \mu) - 1)/(\lambda - \mu)$. □

Since a branching process with immigration does not become extinct, one would expect to be able to estimate the mean and the variance consistently. In the subcritical case we have a stationary and ergodic Markov chain, which can be analyzed either using the work in inference for Markov chain by Billingsley (1962) or using time series methods (Venkataraman, 1982). The supercritical case is similar to the regular BGW situation. Winnicki (1988) tried to unify the theory by introducing conditional least squares estimators. Let $\epsilon_n = Z_n - mZ_{n-1} - \kappa$. Then (ϵ_n) is an $\mathcal{F}_n$-martingale difference sequence with

$$\mathbf{E}(\epsilon_n^2 | \mathcal{F}_{n-1}) = \sigma^2 Z_{n-1} + \beta^2\,.$$

In order to construct appropriately weighted conditional least squares estimators of m and κ we need to minimize $\Sigma_1^n \mathbf{E}(\epsilon_k^2 | \mathcal{F}_{k-1})^{-1} \epsilon_k^2$. However, this criterion depends on the unknown parameters σ and β. Let $\delta_n = \epsilon_n (Z_{n-1} + 1)^{-1/2}$. Then (δ_n) is also an $\mathcal{F}_n$-martingale difference sequence, with

$$\mathbf{E}(\delta_n^2 | \mathcal{F}_{n-1}) = \frac{\sigma^2 Z_{n-1} + \beta^2}{Z_{n-1} + 1}$$

so

$$\min(\sigma^2, \beta^2) \le \mathbf{E}(\delta_n^2 | \mathcal{F}_{n-1}) \le \max(\sigma^2, \beta^2)\,.$$

Here (δ_n) is nearly a constant variance martingale difference sequence, and conditional least squares estimators can be constructed by minimizing $\Sigma_1^n \delta_k^2$, yielding

$$\tilde{m}_n = \frac{\sum_{i=1}^n Z_i \sum_{i=1}^n (Z_{i-1} + 1)^{-1} - n \sum_{i=1}^n Z_i (Z_{i-1} + 1)^{-1}}{\sum_{i=1}^n (Z_{i-1} + 1) \sum_{i=1}^n (Z_{i-1} + 1)^{-1} - n^2}$$

and

$$\tilde{\kappa}_n = \frac{\sum_{i=1}^{n} Z_{i-1} \sum_{i=1}^{n} Z_i (Z_{i-1}+1)^{-1} - \sum_{i=1}^{n} Z_i \sum_{i=1}^{n} Z_{i-1}(Z_{i-1}+1)^{-1}}{\sum_{i=1}^{n} (Z_{i-1}+1) \sum_{i=1}^{n} (Z_{i-1}+1)^{-1} - n^2}.$$

In order to estimate the variances, write

$$\eta_n = \frac{\epsilon_n^2}{Z_{n-1}+1} - \sigma^2 - \frac{\beta^2 - \sigma^2}{Z_{n-1}+1}.$$

Then $\mathbf{E}(\eta_n | \mathcal{F}_{n-1}) = 0$ and

$$\mathbf{E}(\eta_n^2 | \mathcal{F}_{n-1}) = \frac{R(Z_{n-1})}{(Z_{n-1}+1)^2},$$

where $R(x) = 2\sigma^4 x^2 + (\mu_4^P + 4\sigma^2\beta^2 - 3\sigma^4)x + \mu_4^Q - \beta^4$, μ_4^P is the fourth central moment of P, and μ_4^Q that of Q. Clearly, $\mathbf{E}(\eta_n^2 | \mathcal{F}_{n-1})$ is bounded, and we can minimize $\Sigma_1^n \eta_k^2$ to get estimators of σ^2 and β^2, namely

$$\tilde{\sigma}_n^2 = \frac{\sum_{k=1}^{n} (Z_{k-1}+1)^{-2} \sum_{k=1}^{n} \frac{\tilde{\epsilon}_k^2}{Z_{k-1}+1} - \sum_{k=1}^{n} (Z_{k-1}+1)^{-1} \sum_{k=1}^{n} \frac{\tilde{\epsilon}_k^2}{(Z_{k-1}+1)^2}}{n \sum_{k=1}^{n} (Z_{k-1}+1)^{-2} - \left(\sum_{k=1}^{n} (Z_{k-1}+1)^{-1}\right)^2}$$

and

$$\tilde{\beta}_n^2 = \frac{\sum_{k=1}^{n} \frac{Z_{k-1}}{Z_{k-1}+1} \sum_{k=1}^{n} \frac{\tilde{\epsilon}_k^2}{(Z_{k-1}+1)^2} - \sum_{k=1}^{n} \frac{Z_{k-1}}{(Z_{k-1}+1)^2} \sum_{k=1}^{n} \frac{\tilde{\epsilon}_k^2}{Z_{k-1}+1}}{n \sum_{k=1}^{n} (Z_{k-1}+1)^{-2} - \left(\sum_{k=1}^{n} (Z_{k-1}+1)^{-1}\right)^2},$$

where $\tilde{\epsilon}_k = Z_k - \tilde{m}_n Z_{k-1} - \tilde{\kappa}_n$.

The statistical theory of these estimators when $m \leq 1$ does not fall under the scope of our presentation since the model then is ergodic, so in what follows we will only give proofs for the case $m > 1$. The papers by Winnicki (1988) and Wei and Winnicki (1990) detail the subcritical and the (very difficult) critical cases. We first need a technical result.

Lemma 5.1: For $m > 1$

(i) $\Sigma (Z_{k-1}+1)^{-\alpha} \overset{\text{a.s.}}{\to} V^{-\alpha}(m-1)^{-1}$ for all $\alpha > 0$.

(ii) $\Sigma \epsilon_k (Z_{k-1}+1)^{-1} < \infty$ a.s.

(iii) $\Sigma \epsilon_i = O(\log n (\Sigma (Z_{i-1}+1))^{1/2})$ a.s.

Proof: (i) is immediate from Proposition 5.1(i) and (ii), (ii) follows from the martingale strong law (Lemma 2.5). (iii) is from Heyde and Leslie (1971). □

Theorem 5.1: Assuming that $\sigma^2 < \infty$ and $\beta^2 < \infty$,

(i) $\tilde{m} \to m$ almost surely for all $m \neq 1$ and in probability for $m = 1$.

(ii) $\tilde{\kappa} \to \kappa$ almost surely for $m < 1$ and in probability for $m = 1$. For $m > 1$ there is no consistent estimator of κ.

(iii) $(\Sigma X_{i-1} + 1)^{1/2}(\tilde{m}_n - m)$ converges in distribution for all values of m. If $m \neq 1$ the limit law is normal, whereas if $m = 1$ the limit law is that of $(Y(1) - \kappa)(\int_0^1 Y(t)\,dt)^{-1/2}$, where $Y(t)$ is a diffusion process with generator $Af(x) = \frac{1}{2}\sigma^2 x f''(x) + \kappa f'(x)$. The limiting variance when $m > 1$ is σ^2, while when $m < 1$ it is the $(1, 1)$-element of $V^{-1}W(V^{-1})^T$ where

$$V = \begin{pmatrix} \mathbf{E}Z & 1 \\ \mathbf{E}Z/(Z+1) & \mathbf{E}1/(Z+1) \end{pmatrix}$$

and

$$W = \begin{pmatrix} \mathbf{E}\sigma^2 Z + \beta^2 & \mathbf{E}(\sigma^2 Z + \beta^2)/(Z+1) \\ \mathbf{E}(\sigma^2 Z + \beta^2)/(Z+1) & \mathbf{E}(\sigma^2 Z + \beta^2)/(Z+1)^2 \end{pmatrix},$$

and Z has the stationary distribution π.

If, in addition, μ_4^P and μ_4^Q are finite,

(iv) $\tilde{\sigma}_n^2 \to \sigma^2$ almost surely if $m \neq 1$ and in probability if $m = 1$.

(v) $\tilde{\beta}_n^2 \to \beta^2$ almost surely if $m < 1$, in probability if $m = 1$ and $2\kappa \leq \sigma^2$. If $m = 1$ and $2\kappa > \sigma^2$, or if $m > 1$, then β is not estimable.

Proof for $m > 1$: The consistency of $\tilde{m}$ is immediate from Lemma 5.1. The prove the distributional result for $\tilde{m}$, we note that

$$\left(\sum (Z_{i-1} + 1)\right)^{1/2}(\tilde{m}_n - m) = \frac{A_n - B_n}{C_n},$$

where

$$A_n = \sum \epsilon_i \Big/ \left(\sum (Z_{i-1} + 1)\right)^{1/2},$$

$$B_n = \frac{n \sum \epsilon_i/(Z_{i-1} + 1)}{\sum (Z_{i-1} + 1)^{-1} \left(\sum (Z_{i-1} + 1)\right)^{1/2}},$$

and

$$C_n = 1 - n^2 \Big/ \left(\sum (Z_{i-1} + 1) \sum (Z_{i-1} + 1)^{-1}\right).$$

From Lemma 5.1 it follows that $C_n \to 0$ a.s., and that $B_n \to 0$ a.s. For A_n we use the martingale central limit theorem (Lemma 3.1).

To show the consistency of $\tilde{\sigma}_n^2$, let $\bar{\sigma}_n^2$ be $\tilde{\sigma}_n^2$ with ϵ_k replacing $\tilde{\epsilon}_k$. Then

$$\bar{\sigma}_n^2 - \sigma^2 = \frac{D_n - E_n}{1 - F_n},$$

where

$$D_n = \frac{1}{n} \sum \frac{\eta_k}{Z_{k-1} + 1},$$

$$E_n = \frac{\sum (Z_{k-1} + 1)^{-1} \sum \eta_k (Z_{k-1} + 1)^{-2}}{n \sum (Z_{k-1} + 1)^{-2}},$$

and

$$F_n = \left(\sum (Z_{k-1} + 1)^{-1}\right)^2 \Big/ \left(n \sum (Z_{k-1} + 1)^{-2}\right).$$

By the L_2-martingale convergence theorem $\Sigma\, \eta_k/(Z_{k-1} + 1)^2$ converges a.s., whence $B_n \to 0$ a.s. using Lemma 5.1. Similarly $F_n \to 0$ a.s. For D_n, the martingale strong law shows that $D_n \to 0$ a.s. Hence $\bar{\sigma}_n^2 \to \sigma^2$ a.s. It remains to show that $\tilde{\sigma}_n^2 - \bar{\sigma}_n^2 \to 0$ a.s. Write

$$\tilde{\sigma}_n^2 - \bar{\sigma}_n^2 = \frac{G_n - H_n}{1 - F_n},$$

where

$$G_n = \frac{1}{n} \sum \frac{\tilde{\epsilon}_k^2 - \epsilon_k^2}{Z_{k-1} + 1}$$

and

$$H_n = \frac{\sum (Z_{k-1} + 1)^{-1} \sum \dfrac{\tilde{\epsilon}_k^2 - \epsilon_k^2}{Z_{k-1} + 1}}{n \sum (Z_{k-1} + 1)^{-2}}.$$

From above we have that

$$\tilde{m}_n - m = O\left(\sum \epsilon_i/(Z_{i-1} + 1)^{1/2}\right) = O\left(\log n \left(\sum Z_{i-1} + 1\right)^{-1/2}\right) \text{ a.s.}$$

using Lemma 5.1(iii). Furthermore, by somewhat tedious algebra, we can write

$$\tilde{\kappa}_n - \kappa = O\left(\sum \frac{\epsilon_i}{Z_{i-1} + 1} \Big/ \sum (Z_{i-1} + 1)^{-1}\right) \text{ a.s.,}$$

which is finite a.s. by Lemma 5.1(i) and (ii). Now write

$$nG_n = 2(\tilde{m}_n - m)\sum \frac{Z_{k-1}\epsilon_k}{Z_{k-1}+1} + 2(\tilde{\kappa}_n - \kappa)\sum \frac{\epsilon_k}{Z_{k-1}+1}$$
$$+ (\tilde{m}_n - m)[(\tilde{m}_n - m) + (\tilde{\kappa}_n - \kappa)]\sum \frac{Z_{k-1}}{Z_{k-1}+1}$$
$$+ (\tilde{\kappa}_n - \kappa)[(\tilde{m}_n - m) + (\tilde{\kappa}_n - \kappa)]\sum (Z_{k-1}+1)^{-1} .$$

Hence $nG_n = O(\log^2 n)$ a.s., whence $G_n \to 0$ a.s. A similar expansion for H_n shows that $H_n \to 0$ a.s.

Finally, to show the inconsistency of $\tilde{\kappa}$ and $\tilde{\beta}$, we apply a result very similar to Proposition 1.4. A local central limit theorem for the transition density yields, assuming that the lattice size of the immigration distribution is a multiple of that of the offspring distribution, and that the offspring distribution has three moments, that

$$\rho(z_{n+1}|z_n) = s(2\pi\sigma^2 z_n)^{-1/2}\exp(-\tfrac{1}{2}(z_{n+1} - mz_n)^2/\sigma^2 z_n)$$
$$\times (1 + O(nm^{-n/2})) \text{ a.s.,}$$

where s is the lattice size of the offspring distribution. The computations in the proof of Proposition 1.4 are easily adapted to show that the only parameters that can possibly be consistently estimated are the offspring mean, variance and lattice size, and the lattice offsets of both distributions. □

Application: Brownian Motion

Suppose that we observe under a microscope a small, well defined volume of colloidal solution, and count the number of particles in the volume at discrete time intervals $\Delta, 2\Delta, \ldots$. These numbers will be dependent, with the strength of dependence decreasing as Δ increases. Assuming that the motions of individual particles are independent, and that the process is in a stationary state, one can model this process by a death process with immigration. In the example above, the corresponds to letting $\lambda \to 0$, which yields a branching process with immigration with offspring distribution having pgf $1 - \exp(-\mu\Delta) + \exp(-\mu\Delta)s$, and immigration distribution having pgf $\exp(-\nu(1 - \exp(-\mu\Delta))(s-1)/\mu\Delta)$. The stationary distribution of this process is Poisson with mean ν/μ, as is seen from (3) by letting $t \to \infty$ and $\lambda \to 0$. Using data from Experiment C of Westgren (1916), studying a gold hydrosol, we get the following table.

Count	0	1	2	3	4	5	6	7
Observed	381	569	359	176	67	28	5	2
Expected	380	543	388	185	66	19	4	1

The mean of the observed distribution is 1.43, and the chi-squared statistic 8.7. Following Bartlett (1966, Section 8.2) the expectation under the null hypothesis of this statistic is 10.78, with a variance of 57.17. Thus the agreement is good. The data (given in the Appendix) are slightly at variance with Westgren's tabulation, but are based on his listing of the data. The differences are unimportant. Using the weighted least squares estimators developed above, we get $\tilde{\kappa} = 0.555$ and $\tilde{m} = 0.613$. Since the offspring distribution in Bernoulli $(\exp(-\mu t))$, where the interval between observations is $t = 1.39$ s, we estimate μ by $-\log(m)/\Delta = 0.352$, and ν by μ times the stationary mean, or 0.501.

The gold particles in the solution undergo Brownian motion, the theory of which was first presented by Einstein (1905), and further developed in particular by Smoluchowski. Westgren's experiment was a striking confirmation of the Einstein–Smoluchowski theory. To relate Brownian motion computations to the model just discussed, Smoluchowski (1916) computed the probability that a particle present in the observed volume at a given instant would be gone t time units later. This probability, which is $1 - m$, was termed *Wahrscheinlichkeitsnachwirkung*, or probability after-effect. (Chandrasekhar, 1954, is an excellent exposé of the theory involved). Denoting the probability after-effect by P we have that

$$P = 2(1 - \Phi(\beta)) + \frac{2}{\beta(2\pi)^{1/2}}\left(1 - \exp\left(-\frac{\beta^2}{2}\right)\right), \tag{4}$$

where Φ is the normal cdf, and $\beta = h(2Dt)^{-1/2}$, where D is the diffusion coefficient of the particles and h is the thickness of the volume studied. In this experiment $D = 3.95 \times 10^{-8}\ \text{cm}^2\ \text{s}^{-1}$ (we follow Westgren in giving the measurements in the old cgs-units), and $h = 6.56 \times 10^{-4}$ cm. The theoretical calculation yields $1 - m = 0.394$. The asymptotic standard error of the estimate $1 - \tilde{m} = 0.387$ amounts to 0.037, so the estimate conforms well to theory.

The diffusion coefficient D about was actually computed from the formula

$$D = \frac{RT}{N}\frac{1}{6\pi\mu a}$$

where R is the gas constant, T is the absolute temperature, N is Avogadro's number (the number of particles in one mole of a substance), μ is viscosity of the medium, and a is the size of a particle. At the time of Westgren's experiment, the value of Avogadro's number was not well determined (see footnote 28 by Fürth in Einstein, 1956). In fact, one of Westgren's goals in this research was the determination of this important constant. To compute the diffusion coefficient given above, Westgren used $N = 6.06 \times 10^{23}\ \text{mol}^{-1}$. We can use the asymptotic confidence integral obtained from the standard error for $\tilde{m}$ given above to get a confidence interval for N, based on this experiment. Note that

$$N = \frac{\beta^2 RTt}{h^2 3\pi a\mu},$$

where β depends on P through relation (4). The values of these quantities were $R = 8.31 \times 10^7$ g cm^2 s^{-2} mol^{-1} K^{-1}, $T = 290.0$K, $a = 4.95 \times 10^{-8}$ cm, and $\mu = 1.08 \times 10^{-2}$ g cm^{-1} s^{-1}. The confidence interval (0.313, 0.461) for P translates into (1.64, 2.54) for β and $(4.15, 10.0) \times 10^{23}$ for N. The point estimate $1 - \tilde{m}$ corresponds to $N = 6.29 \times 10^{23}$ mol^{-1}. For comparison, the current estimated value of N is 6.02252×10^{23} mol^{-1}. □

2. BRANCHING PROCESSES WITH VARYING ENVIRONMENT

The environment of a branching process is its offspring distribution. So far we have concentrated on constant environment, i.e., the same offspring distribution in each generation. Suppose now that the offspring in generation k follows the offspring distribution (p_{ki}). Then $m_n = \mathbf{E}Z_n = \Pi_0^{n-1}\ \mu_j$, where μ_j is the mean of the offspring distribution for generation j. It is possible to obtain a limit other than extinction or explosion, but only by in essence eventually making the process deterministic.

Proposition 5.2: $Z_n \to Z_\infty$ a.s. where $\mathbf{E}Z_\infty \le \liminf m_n$. Provided that all $p_{k0} < 1$, the limiting r.v. satisfies $\mathbf{P}(Z_\infty = 0) + \mathbf{P}(Z_\infty = \infty) < 1$ iff $\Sigma_0^\infty (1 - p_{n1}) < \infty$.

The long proof of this result is in Jagers (1975, pp. 72–76).

Application: The Spread of Rumors

In the early 1950s, the Washington Public Opinion Laboratory in Seattle carried out "Project Reverse" which was intended to study the diffusion of information (in particular, since the project was funded by the US Air Force, information contained in leaflets dropped from the air). A subexperiment took place in a village with 210 housewives. 42 of these were told a slogan about a particular brand of coffee. Each housewife was asked to pass the information on. As an incentive, participants were told that they would get a free pound of coffee if they knew the slogan when an interview called two days later. It was possible to trace the route by which each hearer had obtained the slogan, so that they could be classified by generations. The data, quoted by Bartholomew (1967), are given below.

Generation	1	2	3	4	5
Size	69	53	14	2	4

In this situation we need to take the small population into account. Only one third of the population had not heard the slogan at the end of the experiment. To model this changing potential population, we may assume that the offspring distribution is size-dependent, having mean $\mu_i = \mathbf{E}(X_{i,k}) = \mu \max(1 - Y_{i-1}/N, 0)$, where N is the initial population at risk. Poisson offspring considerations suggest using $\sigma_i^2 = \mu_i$. Notice that this environment is neither iid nor stationary, so the theory developed below does not apply. If we write the process

$$Z_i = \mu g_i Z_{i-1} + W_i ,$$

where $g_i = (1 - Y_{i-1}/N)$, $\mathbf{E}W_i = \mathbf{Cov}(W_i, W_{i+j}) = 0$, and $\mathbf{Var}\, W_i = Z_{i-1}\sigma_i^2$, we can estimate μ using least squares to get

$$\hat{\mu} = \frac{\sum_1^n Z_i}{\sum_1^n g_i Z_{i-1}} = 2.02 .$$

In other words, in the original generation each housewife told two others on average, while in the final generation the mean rate of spread was 0.29.

□

An immediate generalization of the varying environments is to environments that are somehow selected at random. For example, Sykes (1969) suggests that random environments occur naturally in human population models (see also Keiding, 1975b, 1976). Let (π_n) be a random sequence of offspring distributions, the *environmental process*, and given π_n let offspring of generation n be generated according to π_n. Since conditionally a random environment process reduces to a varying environment process, it is immediate from Proposition 5.2 that

$$\mathbf{P}(Z_\infty = 0 \text{ or } \infty) = 1 \quad \text{iff} \quad \mathbf{P}\left(\text{some } \pi_{n0} = 1 \text{ or } \sum_{n=1}^{\infty} (1 - \pi_{n1}) = \infty\right) = 1 ,$$

where π_{nk} is the probability of k offspring in the nth environment. To verify this criterion we can use the Kolmogorov three series theorem if the environments are iid, or an ergodic theorem if the environmments are stationary. Furthermore, Proposition 5.2 also yields that $\mathbf{E}Z_\infty \le \liminf \mathbf{E}m_n$, so that Z_∞ is a.s. finite provided that $\mathbf{P}(\liminf \Sigma_1^n \log \mu_k < \infty) = 1$. In the stationary ergodic environments case, a necessary and sufficient condition for this is that $\nu = \mathbf{E} \log \mu_1 \le 0$. As a consequence, a random environment process with $\nu > 0$ is called supercritical.

Let $\mathcal{F}_n$ be the σ-algebra generated by $\pi_0, \ldots, \pi_n, Z_0, \ldots, Z_n$. Then

$$\mathbf{E}(s^{Z_{n+1}} | \mathcal{F}_n) = P_n(s)^{Z_n} \text{ a.s.} ,$$

where $P_n(s) = \Sigma\, s^i \pi_{ni}$. It follows, much as in Theorem 1.1 (Athreya and Karlin, 1971) that

$$W_n = Z_n / m_n \to W \text{ a.s.},$$

where $\mathbf{E}(W | \mathscr{F}_\pi) = 1$, and

$$\mathbf{P}(W = 0 | \mathscr{F}_\pi) = q(\pi) = \mathbf{P}(Z_n \to 0 | \mathscr{F}_\pi) < 1 ,$$

where $\mathscr{F}_\pi$ is the σ-algebra generated by $\pi = (\pi_0, \pi_1, \ldots)$. Thus $(W = 0) = (Z_n \to 0)$ a.s., just as in the case of regular BGW processes. By taking expectations, these events have probability $q = \mathbf{E}q(\pi)$.

Proposition 5.3: Assume that the environment process is iid with $\nu > 0$ and $0, \sigma_\nu^2 = \mathbf{Var} \log \mu_1 < \infty$. Then

$$(Z_n \exp(-n\nu))^{n^{-1/2}} \xrightarrow{d} Ue^V ,$$

where U and V are independent with $U = 1(Z_n \to \infty)$ and $V \sim \mathbf{N}(0, \sigma_\nu^2)$.

Proof: Let $Y_n = m_n \exp(-n\nu)$, so that

$$Z_n e^{-n\nu} = W_n Y_n .$$

For a sequence (k_n) of integers with $k_n \to \infty$ and $k_n n^{-1/2} \to 0$, let $U_n = 1(Z_{k_n} > 0)$ and $X_n = \mu_{k_n} \cdots \mu_{n-1} \exp(-n\nu)$. Since U_n depends on $(\pi_0, \ldots, \pi_{k_n - 1}, Z_0, \ldots, Z_{k_n})$ only, U_n and X_n are independent. Since $W_n \to W$ a.s., we have $W_n \equiv 0$ for all n large enough on the set $(W = 0)$. On $(W > 0)$ we have $W_n^{n^{-1/2} n} \to 1$ a.s. Since $(W = 0) = (Z_n \to 0)$ a.s., we deduce that $U_n - W_n^{n^{-1/2}} \to 0$ a.s.

Next note that

$$n^{-1/2} \log Y_n = \frac{1}{n^{1/2}} \left(\sum_1^n \log \mu_i - n\nu \right) \xrightarrow{d} \mathbf{N}(0, \sigma_\nu^2)$$

by the central limit theorem. It remains to show that $Y_n^{n^{-1/2}} - X_n^{n^{-1/2}} \xrightarrow{\mathbf{P}} 0$. But

$$Y_n^{n^{-1/2}} - X_n^{n^{-1/2}} = (m_n e^{-n\nu})^{n^{-1/2}} (1 - m_{k_n}^{-n^{-1/2}}) .$$

The first term on the right-hand side converges in probability to 1. By the central limit theorem, $n^{-1/2} \log m_{k_n} = n^{-1/2}(k_n \nu + k_n^{1/2} O_{\mathbf{P}}(1))$ which goes to zero by the choice of k_n, whence $m_{k_n}^{n^{-1/2}} \xrightarrow{\mathbf{P}} 1$. □

In other words, Z_n, given nonextinction, is asymptotically lognormal with median $\exp(n\nu)$. This distribution is completely determined by the environmental fluctuations. It is generally assumed in ecology that for large populations the environmental fluctuations are more important than sampling variability. However, the variable U governing the event of extinction is determined by the sampling variations given the environment.

The statistical theory of random environment branching processes is not very well developed. The following result is central.

Proposition 5.4: Let (Z_k) be a branching process with iid environments (π_i). The empirical distribution function of the random variables $\bar{m}_k = Z_k / Z_{k-1}$, $k = 1, 2, \ldots$ converges a.s. on the explosion set to the distribution function of μ_1.

Proof: Given π_1, let ξ_k, $k = 1, 2, \ldots$ be iid according to π_1. Then by the law of large numbers

$$1\left(\frac{1}{n_k} \sum_{k=1}^{n_k} \xi_k \le x\right) \xrightarrow{d} 1(\mu_k \le x)$$

as $n_k \to \infty$. Thus

$$\mathbf{E}(|1(Z_{k+1} \le xZ_k) - 1(\mu_k \le x)| / Z_k) \le f(Z_k) \to 0 \qquad \text{as } Z_k \to \infty$$

whenever x is a continuity point for the distribution of μ_k. Hence $(1/n) \times \Sigma_k f(Z_k) \to 0$ on $Z_k \to 0$. □

Corollary: Let

$$\tilde{m} = \frac{1}{n} \sum_{1}^{n-1} Z_{k+1}/Z_k ,$$

$$\tilde{\sigma}_\mu^2 = \frac{1}{n} \sum_{1}^{n-1} (Z_{k+1}/Z_k - \tilde{m}_n)^2 ,$$

$$\tilde{\nu} = \frac{1}{n} \sum_{1}^{n-1} \log(Z_{k+1}/Z_k) = \log Z_n/n ,$$

and

$$\tilde{\sigma}_\nu^2 = \frac{1}{n} \sum_{1}^{n-1} (\log(Z_{k+1}/Z_k) - \tilde{\nu})^2 .$$

Then $\tilde{m} \to \mathbf{E}\mu_1$, $\tilde{\nu} \to \nu$, $\tilde{\sigma}_\mu^2 \to \mathbf{Var}\ \mu_1$, and $\tilde{\sigma}^2{}_\nu \to \mathbf{Var} \log \mu_1$ in probability on the explosion set. Furthermore,

$$n^{1/2}(\tilde{m} - m) \xrightarrow{d} \mathbf{N}(0, \mathbf{Var}\ \mu_1)$$

and

$$n^{1/2}(\tilde{\nu} - \nu) \xrightarrow{d} \mathbf{N}(0, \mathbf{Var} \log \mu_1) .$$

The asymptotic variances can be estimated by $\tilde{\sigma}_\mu^2$ and $\tilde{\sigma}_\nu^2$, respectively.

Proof: The consistency follows from Proposition 5.4. Asymptotic normality is a consequence of Proposition 5.3. □

Remark: The assumption of iid environments can be weakened to stationarity, provided some mixing assumption holds (Hall and Heyde, 1980, section 5.3, contains appropriate results for such a generalization). It is not known what parameters of a random environment branching process are consistently estimable, conditional upon nonextinction.

3. BRANCHING PROCESSES WITH CONTINUOUS STATE SPACE

A stochastic process $(X(t), t \geq 0)$ is called a *subordinator* if it is nonnegative, has stationary and independent increments, and paths that are right continuous with left limits and $X(0)=0$. Clearly, $X(t)$ must have an infinitely divisible distribution. We use $X(t)$ to define a continuous state space branching process $(Z_n, n=0,1,\ldots)$, by

$$Z_{n+1} = X(S_n) - X(S_{n-1}) ,$$

where $Z_0 = 1$, $S_{-1} = 0$, and $S_n = \Sigma_0^n Z_k$. This is very similar to the construction of a discrete state space branching process using a random walk (Exercise 1.6). The distribution of $X(1)$ is called the offspring distribution, and assumed to have cumulant generating function $h(s) = -\log \mathbf{E}(\exp(-sX(1)))$. The basic branching property holds:

$$\mathbf{E}(\exp(-sZ_{n+1}) | \mathcal{F}_n^Z) = \exp(-h(s)Z_n) \qquad (5)$$

almost surely. Continuous state space branching processes behave much like their discrete state space counterparts.

Proposition 5.5

(i) $q = \lim \mathbf{P}(Z_n = 0) = 0$ iff $\mathbf{P}(Z_1 = 0) = 0$. If $\mathbf{P}(Z_1 = 0) > 0$, then q is the largest solution to the equation $h(s) = s$.
(ii) Let $h_n(s) = -\log \mathbf{E}(\exp(-sZ_n))$. Then $\lim h_n(s) = -\log \mathbf{P}(\lim Z_n = 0)$.
(iii) $\mathbf{P}(\lim Z_n = 0) = 1 - \mathbf{P}(\lim Z_n \to \infty)$.
(iv) Assume that $\mathbf{Var}\, Z_1 < \infty$. Let $W_n = m^{-n} Z_n$, where $m = \mathbf{E}Z_1$. Then $W = \lim W_n$ exists a.s. with $W = 0$ a.s. if $m \leq 1$, and $\mathbf{E}W = z_0$ if $m > 1$.

Proof: If $q_n = \mathbf{P}(Z_n = 0)$ we have $0 \leq q_n \uparrow q$. Hence $q = 0$ implies that $q_1 = 0$. Conversely, if $q_1 = 0$,

$$\lim_{s \to \infty} h(s) = \lim_{s \to \infty} -\log \mathbf{E}(\exp(-sX(1)) = -\log \mathbf{P}(Z_1 = 0) = \infty$$

by dominated convergence. Therefore $-\log \mathbf{P}(Z_n = 0) = \lim_{s\to\infty} h_n(s) = \infty$, so $q = \lim q_n = 0$. If $q_1 > 0$ we have $q_n > 0$ for all n, so $h_n(\infty) < \infty$. Thus

$$h(-\log q) + \log q = \lim(h(h_n(\infty)) - h_n(\infty)) = \lim(-\log q_{n+1} + \log q_n) = 0 .$$

The maximality of the solution is easy.

To verify (ii), first write

$$h_n(s) \le -\log \mathbf{E} \exp(-sZ_n)1(\lim Z_n = 0) . \tag{6}$$

As $n \to \infty$ the right-hand side tends to $-\log \mathbf{P}(\lim Z_n = 0)$. If $h(s) < s$ for all s we have $m = \mathbf{E}Z_1 \le 1$. Hence (Z_n) is a nonnegative supermartingale, so it converges to a finite limit. But $h_n(s) \to 0$ as $n \to \infty$, so the limit must be zero a.s. If $h(s) > s$ for all s, we have $h_n(s) \to \infty$ as $n \to \infty$, so by (6) $-\log \mathbf{P}(\lim Z_n = 0) = \infty$. If neither of these conditions on h hold it is easy to see that there must be a unique s_0 such that $h(s_0) = s_0$, and $\lim_{n\to\infty} h_n(s) = s_0$. Now $\exp(-s_0 Z_n)$ is a bounded martingale by (5), and hence converges a.s. to some r.v. Y_∞ with

$$-\log \mathbf{E} Y_\infty^{s/s_0} = \lim h_n(s) = s_0 1(s > 0) ,$$

whence $\mathbf{P}(\lim Z_n = 0) = \mathbf{P}(Y_\infty = 1) = 1 - \exp(-s_0)$.

From (ii) and the convergence theorem for Laplace transforms it follows that $Z_n \to Z$ in distribution, where $Z = 0$ or ∞ with probability 1. From the proof of (ii) it follows that if $\mathbf{P}(Z = 0) > 0$ the convergence is a.s. On the other hand, if $\mathbf{P}(Z = 0) = 0$ we have $h(s) > s$, so that $\mathbf{P}(Z_1 \ge 1) = 1$, and therefore (Z_n) is a nondecreasing sequence, which converges a.s. In either case the a.s. limit must have the same distribution as Z.

Finally, (iv) follows from the branching equation (5) and the martingale convergence theorem. □

The statistical theory for continuous state space branching processes is very similar to the BGW case. We start with the estimator

$$\bar{m}_n = \frac{Z_n}{Z_{n-1}} \stackrel{d}{=} \frac{X(Z_{n-1})}{Z_{n-1}}$$

since $Z_n = X(S_{n-1}) - X(S_{n-2}) \stackrel{d}{=} X(S_{n-1} - S_{n-2}) = X(Z_{n-1})$. The conditional mean, given $Z_{n-1} > 0$, is

$$\mathbf{E}(\bar{m}_n | Z_{n-1} > 0) = \mathbf{E}X(1)$$

using the fact that $\mathbf{E}X(at) = a\mathbf{E}X(t)$ which is immediate from the stationarity of the increments of X. We use Proposition 5.3(iv) to see that $\bar{m}_n \to m$ a.s. on the explosion set.

As in the BGW case, we can utilize the observations better by looking at a weighted average of the $\bar{m}_i$. Since $\mathbf{Var}(\bar{m}_n | Z_{n-1}) = Z_{n-1}\ \mathbf{Var}\ X(1)$ we use Z_{i-1}/S_{n-1} as weights. This yields $\hat{m}_n = (S_n - z_0)/S_{n-1}$. Since $S_n = X(S_{n-1})$, we can write this

$$\hat{m}_n = \frac{X(S_{n-1}) - z_0}{S_{n-1}} .$$

Since $X(S_{n-1}) - S_{n-1}m$ is an $\mathcal{F}_n^Z$-martingale, Lemma 2.5 applies to show that $\hat{m}_n$ is consistent on the explosion set $\{Z_n \to \infty\}$.

Example: The Gamma Process

The most common example of a subordinator is the gamma process, for which $X(t) \sim \Gamma(t, 1/m)$. We observe $z_1, \ldots, z_n$, or equivalently $X(s_0) - X(s_{-1}), \ldots, X(s_{n-1}) - X(s_{n-2})$. But given $\mathcal{F}_i^S$ we have that

$$X(s_i) - X(s_{i-1}) \sim \Gamma(s_i - s_{i-1}, 1/m) ,$$

so the likelihood function based on $z_1, \ldots, z_n$ is proportional to

$$m^{-\Sigma_0^{n-1}(s_i - s_{i-1})}\left(\prod_1^n z_i^{s_i - s_{i-1} - 1}\right)e^{-\Sigma_1^n z_i/m}$$

and the maximum likelihood estimator of m is $\hat{m}$. □

Application: Monitoring Pregnancy Development

In evaluating the results of laboratory tests one often compares the patient to a standard based on data from an appropriate group of healthy subjects. For monitoring purposes much can be gained by using the patient as his or her own reference. This eliminates the intersubject variability from the comparison. In order to implement this kind of procedure one must have baseline measurements on the patient, obtained during a period when the patient was in a stable state. In addition, one must have a model for predicting future observations under the assumption of continued stable condition.

In monitoring the progress of pregnancy, various hormone concentrations can be used to assess the growth of the placenta (Winkel et al., 1976). For example, progesterone is produced in the placenta, and its concentration in blood samples indicates the growth of the placenta, at least after the 12th week of pregnancy. A simple growth model is based on a branching process model, where the number of new placenta cells at time $t+1$ is proportional on average to the total size at time t. The constant of proportionality is the average number of cells produced from one initial cell in one unit of time. If we assume that the concentration of progesterone is a linear function of the

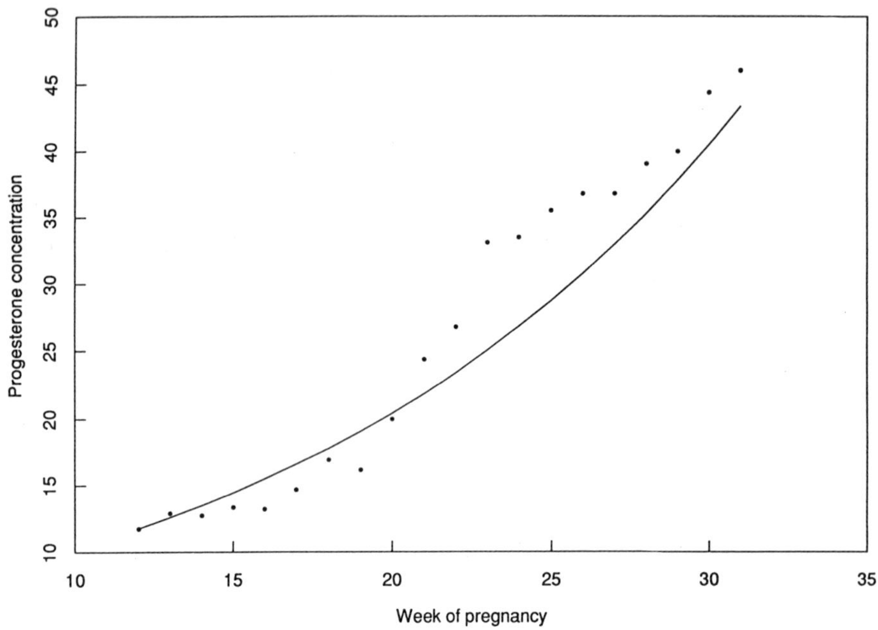

FIGURE 5.1. Progesterone concentration as a function of time for a pregnant woman.

total size of the placenta, we can predict the future growth on the basis of earlier measurements. An example of the path of such measurements for an individual is shown in Figure 5.1.

Assume that the concentration c_n at time n is a continuous state space branching process. If we observe the concentration at times k and $k+1$, we can estimate m by c_{k+1}/c_k. A conditional standard error for this estimator, given the history up to time k, and assuming that the underlying subordinator is a gamma process as in the example above, is

$$\mathbf{Var}(\bar{m}_{k+1}|c_k) = \mathbf{Var}(X(c_k)/c_k) = \mathbf{Var}(X(1))/c_k = m^2/c_k \,.$$

We would predict the next measurement to be $c_{k+1}\bar{m}_{k+1}$, with a conditional prediction variance given c_k of

$$\mathbf{Var}(X^2(c_k)/c_k|\mathcal{F}_k^c) = 2m^4(c_k+1)(2c_k+3)/c_k \,.$$

Applying this to the Winkel data, we find that $\bar{m}_{20} = 1.221$, so the predicted value for week 21 is 24.4 with a standard error of 14.2. Since the actual measurement is 26.8 the process appears to be progressing normally. A more accurate estimator is $\hat{m}$, but the prediction error using this estimator is more difficult to compute, and only worthwhile if we can verify that the

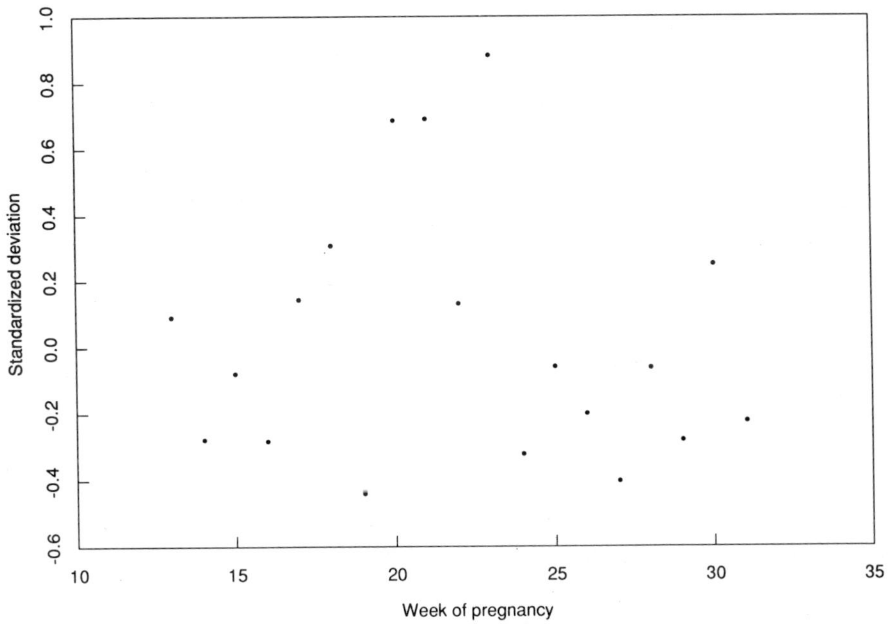

FIGURE 5.2. Standardized deviation of $\bar{m}_k$ from $\hat{m}_n$ for the progesterone data.

assumptions made are reasonable. To assess the gamma process assumption, we must use the entire data set. Here $\hat{m}_n = 1.071$. As in the BGW case we can estimate the variance of $X(1)$ by $\Sigma_{k-1}^{n-1} c_i(\bar{m}_{i+1} - \hat{m}_n)^2/(n-k+1)$. In this case this variance is estimated as 0.144, substantially below the gamma process estimate 1.15. One would therefore expect that the prediction intervals from the gamma process are too wide, perhaps by a factor of three. Figure 5.1 also shows the estimated mean increase of the process (the solid line). There are indications of considerable deviations from the mean. Figure 5.2 shows the differences $|\bar{m}_i - \hat{m}_n|$, normalized by their conditional standard error. This highlights no sizable deviations (when measured in standard units). Going back to the discussion in section 2, a better estimator of m if we had a random environment would be (Exercise 5) $m^* = \Sigma\, \bar{m}_{i+1}/n$. This amounts to 1.077, with the estimated standard deviation of m^* being 0.085. □

4. MULTITYPE PROCESSES

A multitype BGW branching process admits K different types of objects. An individual of type k has offspring of different types according to a K-variate distribution $p_k(x_1, \ldots, x_K)$. We write $Z_n^{(i)}$ for the number of type

i individuals in the nth generation. Let $\mathbf{Z}_n = (Z_n^{(1)}, \ldots, Z_n^{(K)})^T$. Then we can write

$$\mathbf{Z}_{n+1} = \sum_{k=1}^{K} \sum_{j=1}^{Z_n^{(k)}} X_{nj}^{(k)} ,$$

where the $X_{nj}^{(k)}$ represent the number of type j offspring to a type k individual in generation n, and are assumed independent with

$$\mathbf{P}(X_{nj}^{(k)} = x_j , \ j = 1, \ldots, K)) = p_k(x_1, \ldots, x_K) .$$

Let m_{ij} be the mean number of type j offspring for a type i individual, and let $\mathbf{m} = (m_{ij})$. Then starting from a single ancestor at time 0

$$\mathbf{EZ}_n = e(\mathbf{z}_0)\mathbf{m}^n ,$$

where $e(\mathbf{z}_0)$ has zeros everywhere except for a one in the position of the type of $\mathbf{z}_0$. Write $P_k(s) = \Sigma_{\mathbf{x}} \, p_k(\mathbf{x}) \, \Pi_{j=1}^{K} s_j^{x_j}$ and let $P = (P_1, \ldots, P_K)$.

Example: Two Degenerate Cases

Let $P_k(s) = \Sigma_{j=1}^{K} p_{kj}s_j$. This is a case that corresponds to a single individual moving about in a K-state Markov chain. In this situation there is no branching, or, in other words, the offspring distribution is degenerate at 1.

If, on the other hand, $P_k(s) = P(s)$ independent of k, we have the individuals of a branching process moving around according to a K-state Markov chain. □

Application: The Dispersion of a Neutral Allele Among Migrating Flies

Consider individuals of the same species living in colonies located at fixed positions. Suppose that the individuals reproduce and migrate periodically, so that we can consider a population with discrete generations. At some point in time a mutation occurs, producing the mutant allele A_1. The mutants have offspring according to a distribution Q, with mean μ less than one (so that extinction of the allele is certain) and migrate between colonies according to a Markov chain with transition matrix $P = (P_{ij})$. The type of an individual is its colony membership. Consider Y_i, the total number of A_1 genes to appear in colony i. The generating function for Y_i is

$$F_i = s_i Q(\Sigma \, p_{ij} F_j) .$$

We then have

$$m_{ij} = \mu \sum_{k=1}^{K} p_{ik} m_{kj} + 1 \qquad (i = j)$$

so that

$$\mathbf{m} = (I - \mu P)^{-1} ,$$

where I is the identity matrix. Different models of dispersion of individuals and location of colonies yield different structure for P. Suppose for simplicity that the colonies are situated at integer points along an infinite line, and that a new-born mutant stays in the original colony with probability $1 - r$, and moves to a neighboring colony, each with probability $r/2$. Then the solution is

$$m_{ij} = a\lambda^{|j-i|} ,$$

where

$$a = \frac{1}{r\mu(z^2 - 1)^{1/2}} , \qquad \lambda = z - (z^2 - 1)^{1/2} , \qquad z = \frac{1 - \mu(1 - r)}{r\mu} .$$

Now suppose that there is a mutation at colony 0. An immigrant to the nth colony is an individual arriving at the nth colony whose ancestors failed to reach it. Let π_{in} be the probability that i immigrants arrive at the nth colony. Then

$$m_{0n} = m_{00} \sum_{i=1}^{n} i\pi_{in} .$$

Thus we can bound the probability that no individuals will reach colony n by

$$\pi_{0n} \le \sum_{i=1}^{\infty} i\pi_{in} = \lambda^n .$$

We have $\mu = 1 - u$, where u is the mutation rate, typically of the order 10^{-5} to 10^{-6}. A Taylor expansion yields $z \approx (1 + u/r)$ and $\lambda \approx 1 - (2u/r)^{1/2}$. To get an idea of the size of r, we look at data from an experiment carried out by Dobzhansky and Wright (1947), who released a total of 25,134 orange-eyed *Drosophila pseudoobscura* flies from a single point during the summer of 1945. Early in the summer of 1946, flies were collected at the point of release and at traps 500 and 1000 m south and north of this point. The orange gene is recessive, and although homozygous orange flies have not been found in natural populations, heterozygous females occur in different parts of the geographic distribution of these flies. The prevalence of the orange gene prior to the release of the flies was estimated to be 0.2%. No homozygous flies were found in the traps, but some flies were found to be orange heterozygotes. The concentration of orange heterozygotes was decreasing with distance from the point of release, and the standard deviation of the distribution of heterozygotes was estimated between 0.43 and 0.73 km. Thus, the dispersion of the orange gene appears rather slow. An estimate of the dispersion is obtained by assuming a circular normal distribution, and using the higher estimate of the standard deviation. Then 99% of the progeny is contained within a circle of radius 2.2 km. We therefore would assume that colonies are separated by 2.2 km. A naive estimate of the probability of moving from one colony to another is obtained by looking at the dispersion of the flies a week after release, which had a standard deviation of 0.21 km. Using Chebyshev's inequality, we estimate the probability of moving farther than 2.2 km in a week (corresponding very roughly to the lifetime in the field of these flies) to be 0.0089, corresponding to $r = 0.0045$. In order to assess how far any of the mutations may travel, we find an n such that $\pi_{0n} < 0.01$, i.e.,

$$n \approx \log(0.01)/\log(\lambda) \approx 216$$

using a mutation rate of 10^{-6}. In other words, most of the offspring of orange-eyed flies will move no farther than 475 km (or 300 miles). □

The behavior of multitype branching processes has some aspects in common with the behavior of a Markov chain: one has to deal with the possibility of periodicity, of closed classes of types, etc. In order to simplify matters we shall assume that the process is positive regular, i.e., that there is a positive integer n such that all entries in $\mathbf{m}^n$ are strictly positive. Then we can apply Perron–Frobenius theory, and the largest eigenvalue ρ of $\mathbf{m}$ determines the asymptotic behavior of $\mathbf{m}^n$. Extinction is governed by the solution to the equation $P(s) = s$ lying closest to the origin in the unit cube. In the supercritical case, where $\rho > 1$, we have, provided that $\mathbf{E}Z_1^{(i)} \log Z_1^{(i)} < \infty$ for all i), that

$$\rho^{-n}\mathbf{Z}_n \xrightarrow{\text{a.s.}} W\mathbf{v} ,$$

where $\mathbf{v}$ is the left eigenvector of m corresponding to the eigenvalue ρ, and W is a random variable with continuous density except for a jump at the origin.

A natural estimator of m_{ij}, based on observing the number $Y_n^{(ij)}$ of offspring of type j to type i parents, is

$$\hat{m}_{ij} = \frac{Y_n^{(ij)}}{Y_n^{(i)}} ,$$

where $Y_n^{(i)} = \Sigma_{j=1}^K Y_n^{(ij)}$. This is the maximum likelihood estimator based on observing the entire family tree, and using Proposition 2.5 and its Corollary 2, also based on observing only the $Y_n^{(ij)}$. Frequently we would not have information this detailed. The most interesting parameter is often the growth rate ρ. An estimator of ρ can be obtained by ignoring the types. Let

$|\mathbf{Z}_i| = \Sigma_{j=1}^{K} Z_i^{(j)}$ be the total number of individuals in generation i. Estimate ρ by

$$\tilde{\rho} = \frac{\sum_{i=1}^{n} |\mathbf{Z}_i|}{\sum_{i=0}^{n-1} |\mathbf{Z}_i|} .$$

The asymptotic behavior of $\tilde{\rho}$ is more complicated than what we have seen in the other models. For simplicity, we will only discuss the two-type case (Asmussen and Keiding, 1978, contains general theory). Since $\rho^{-n}\mathbf{Z}_n \xrightarrow{\text{a.s.}} W\mathbf{v}$ we have that $\rho^{-n}|\mathbf{Z}_n| \xrightarrow{\text{a.s.}} W|\mathbf{v}|$, and consistency of $\tilde{\rho}$ follows as in Theorem 2.1. To investigate the limiting distribution, write

$$\tilde{\rho} - \rho = \frac{\sum_{k=0}^{n-1} (\mathbf{1}^T\mathbf{Z}_{k+1} - \rho\mathbf{1}^T\mathbf{Z}_k)}{|\mathbf{Z}_0| + \cdots + |\mathbf{Z}_{n-1}|} = \frac{S_n + T_n}{|\mathbf{Z}_0| + \cdots + |\mathbf{Z}_{n-1}|}$$

where

$$S_n = \sum_{0}^{n-1} \mathbf{1}^T(\mathbf{Z}_{k+1} - \mathbf{Z}_k\mathbf{m})$$

and

$$T_n = \sum_{0}^{n-1} \mathbf{d}^T\mathbf{Z}_k \qquad \text{and} \qquad d = \mathbf{1}^T(\mathbf{m} - \rho\mathbf{I}) .$$

Here S_n is a martingale, and an application of Lemma 3.1 shows that

$$(W\rho^n\beta_n^2)^{-1/2}S_n \xrightarrow{d} \mathbf{N}(0, 1)$$

where $\beta_n^2 = \rho^{-1}((1-\rho^{-n})/(1-\rho^{-1}))\,\Sigma_{i=1}^2\, \nu_i\, \mathbf{Var}^i|\mathbf{Z}_1|$, where $\mathbf{Var}^i$ denotes variance for a process starting with one individual in category i.

On the other hand, T_n is a linear functional. The limit theory for such functionals (Kesten and Stigum, 1966; Athreya, 1971; Asmussen, 1976) depends on the relative sizes of the eigenvalues ρ and ρ_1 of $\mathbf{m}$. Let $\mathbf{b}$ be a real vector. Then $\rho^{-n}\mathbf{b}^T\mathbf{Z}_n \xrightarrow{\text{a.s.}} W\mathbf{b}^T\mathbf{v}$, and if $\mathbf{b}^T\mathbf{v} \neq 0$, this settles the asymptotic behavior of T_n. However, if $\mathbf{b}^T\mathbf{v} = 0$ (implying that $|\rho_1| < \rho$), we must distinguish three cases. If $\rho_1^2 > \rho$ we have that $|\rho_1|^{-n}\mathbf{b}^T\mathbf{Z}_n - H_n \xrightarrow{\text{a.s.}} 0$ for some random variable H_n with $\limsup H_n < \infty$ and $\liminf |H_n| > 0$. If $\rho_1^2 = \rho$, $(Wn\rho^n)^{-1/2}\mathbf{b}^T\mathbf{Z}_n$ is asymptotically normally distributed, and if $\rho_1^2 < \rho$, $(W\rho^n)^{-1/2}\mathbf{b}^T\mathbf{Z}_n$ is asymptotically normal. Thus, if $\rho_1^2 < \rho$, S_n and T_n are of the same order of magnitude, while T_n dominates S_n if $\rho_1^2 \geq \rho$.

The following result summarizes the asymptotic behavior of $\tilde{\rho}$ using the theory outlined above.

Theorem 5.2: Let $(\mathbf{Z}_n)$ be a two-type BGW process with $\rho > 1$. The following results are valid on the set of nonextinction.

(i) $\tilde{\rho} \xrightarrow{\text{a.s.}} \rho$.

(ii) Suppose that $\rho_1^2 < \rho$, and $\mathbf{Var}^i|\mathbf{Z}_1| < \infty$, $i = 1, 2$. Then

$$\left(W \sum_0^{n-1} \rho^i\right)^{1/2} (\tilde{\rho}_n - \rho) \xrightarrow{d} \mathbf{N}(0, \sigma_1^2)\,, \qquad \text{for some } 0 < \sigma_1^2 < \infty\,.$$

(iii) Suppose that $\rho_1^2 = \rho$, and $\mathbf{Var}^i|\mathbf{Z}_1| < \infty$, $i = 1, 2$. Then

$$\left(W \sum_0^{n-1} \frac{\rho^i}{n}\right)^{1/2} (\tilde{\rho}_n - \rho) \xrightarrow{d} \mathbf{N}(0, \sigma_2^2)\,, \qquad \text{for some } 0 < \sigma_2^2 < \infty\,.$$

(iv) Suppose that $\rho_1^2 > \rho$. Then $W(\rho/\rho_1)^n(\tilde{\rho}_n - \rho)$ converges a.s.

5. SOURCES

Proposition 5.1 follows Asmussen and Hering (1983). Heyde and Seneta (1972a) first considered estimation in the subcritical branching process with immigration. Bhat and Adke (1981) discussed estimation based on observing the actual number of immigrants at each time. Hudson (1982) presented likelihood ratio tests based on observing individual offspring as well as immigration numbers. Our presentation is based on work by Winnicki (1988). The application follows Chandrasekhar (1954) and Westgren (1916).

Proposition 5.3 is due to Keiding and Nielsen (1973). Becker (1977) discusses applicability of random environments to epidemic processes, and Keiding (1975b) treats biological applications. Proposition 5.4 is standard folklore. Work in this area has been done by Harrington and Fleming (1978), Dion and Esty (1979), Nanthi (1979), and Pakes and Heyde (1982).

Proposition 5.2 comes from Kallenberg (1979), Zhang (1983) analyzed the martingale W_n. The estimation theory is (to my knowledge) new, although not much is different from the discrete state space case. Winkel et al. (1976) analyzed the progesterone data using a diffusion process.

The estimation theory in section 4 is based on work by Asmussen and Keiding (1978) and follows the discussion in Asmussen (1982) and Asmussen and Hering (1983). The estimator $\tilde{\rho}$ was proposed by Becker (1977). Crump and Gillespie (1976) developed the multitype branching process model for the dispersion of mutations in *Drosophila*.

EXERCISES

1. Let Z_n be a branching process with immigration, with offspring mean m and immigration mean κ and finite offspring variance σ^2 and immigration variance β^2. Determine the unweighted conditional least squares estimators of κ of m based on minimizing $\Sigma\, \epsilon_k^2$, and show that the resulting estimator of m is consistent in the supercritical case. What can you say about the estimator of κ?

2. For a branching process with immigration, show that

$$\tilde{m}_n - \sum_{i=1}^{n} Z_i \Big/ \sum_{i=1}^{n} Z_{i-1} = O\left(n^2 \Big/ \Sigma\, (Z_{i-1} + 1)\right) \text{ a.s.}$$

Deduce the asymptotic properties of $\tilde{m}$ by deriving those of $\hat{m} = \Sigma\, Z_i / \Sigma\, Z_{i-1}$ in the supercritical case.

3. Looking at the number of young birds as a function of population size for the whooping crane population (see Appendix) suggests that an immigration-death process may be a better fit than the birth-death model fit in section 2.4.
   (a) Estimate the parameters of the immigration-death model.
   (b) More generally, estimate the parameters of an immigration-birth-death model, and assess the significance of the birth parameter.
   (c) Estimate the extinction probability from the 1972 population size.

4. A simple conceptual model of rainfall in the following: a rain cell is a volume in space containing small droplets. During a short time interval some droplets will leave the rain cell, some new ones will be created, and there is a complex interaction of droplets adjoining and splitting, largely dependent on size. The observed rainfall can, based on this conceptual model, be described as

$$Z_n = \sum_{i=1}^{Z_{n-1}} X_{n,i} + I_n$$

with $m = \mathbf{E}X_{n,i} < 1$. I_n can be thought of as the occurrence of a rain storm, while the $X_{n,i}$ describe the duration and amount of rain (Kedem and Chiu, 1987).
   (a) Assume that $|Z_n - Z_{n-1}| = o(Z_{n-1})$ and that $(1/n)\, \Sigma\, \mathbf{E}((\epsilon_i / Z_{i-1})^2 | \mathcal{F}_{i-1}) \to c^2 > 0$ a.s., where the ϵ_n are defined as in section 1. Show that $n^{-1/2}\, \Sigma\, \epsilon_i / Z_{i-1} \xrightarrow{d} \mathbf{N}(0, c^2)$.

   Remark: The first condition can be satisfied by taking the sampling interval of the (continuous time) rain rate process small enough.

(b) Deduce that

$$\left(\frac{Z_n}{z_0}\right)^{n^{-1/2}}\left(1+(1-m)-\frac{\lambda}{n}\sum_1^n \frac{1}{Z_{i-1}}\right)^{n^{1/2}} \xrightarrow{d} \text{lognormal}(0, c^2)$$

so that, in particular $(Z_n/z_0)^{n^{-1/2}}$ is approximately lognormal$(0, c^2)$ when $m \approx 1$ and $\lambda \approx 0$.

Hint: Let $\delta_n = (Z_n - Z_{n-1})/(Z_{n-1} + 1(Z_{n-1} = 0))$, and show, conditionally upon rain, that $Z_n = (1+\delta_n)\ldots(1+\delta_1)z_0$.

5. Let Z_n be a binary splitting process with varying offspring distribution

$$p_{n0} = 1 - p_n\,, \qquad p_{n2} = p_n\,.$$

Exhibit a sequence p_n such that all $p_n > \frac{1}{2}$ (i.e., all generations are supercritical), but $\mathbf{P}(Z_n \to 0) = 1$.

6. For a random environment branching process with iid environment, find the best linear combination of the $\bar{m}_k$, $k = 1, \ldots, n$, in terms of minimizing the variance of the estimator.

7. Show that $\tilde{\pi}_n = \log \tilde{m}_n$, where $\tilde{m}$ is the estimator given in Exercise 2.3, is asymptotically normal in a process with iid environments, and find its asymptotic efficiency relative to $\tilde{\nu}$ of section 2.

8. Using the Nigerian smallpox epidemic data (see the Appendix), assess the vaccination proportion needed to prevent another outbreak if you take into account the varying size of the population of susceptibles (cf. Exercise 2.17).

9. Let

$$X_{n+1} = X_n + \sum_{k=1}^{X_{n-1}} Y_{n,k}\,,$$

where $Y_{n,k}$ are iid with mean m.

(a) Show that if $m = 1$ then $(\mathbf{E}X_n)$ is the Fibonacci sequence, and X_n/X_{n-1} converges a.s. to the golden ratio $\frac{1}{2}(5^{1/2}+1)$.

(b) Let $Z_0 = (X_0, 0)$, $Z_n = (X_n - X_{n-1}, X_{n-1})$. Show that (Z_n) is a bivariate BGW process. Determine $\mathbf{M}$ and its largest eigenvalue ρ.

(c) Show that $\hat{m} = (X_N - X_1)/\Sigma_0^{N-2} X_n$ is the nonparametric mle of m.

(d) Determine the asymptotic behavior of $\hat{\rho} = \rho(\hat{m})$.

10. Consider a family stemming from a single ancestor. Assuming that each family member mates with a randomly chosen individual from outside the family, and considering the evolution of a particular gene and alleles A and a, we get a multitype process with three types: AA, Aa, and aa.

(a) Provided that we can ignore the sex of an individual and neglect mutation, show that the transition matrix for the process is

$$P = (p_{ij}) = (\mathbf{P}(\text{type } i \text{ individual has type } j \text{ offspring}))$$
$$= \begin{pmatrix} p & q & 0 \\ \frac{1}{2}p & \frac{1}{2} & \frac{1}{2}q \\ 0 & p & q \end{pmatrix}.$$

(b) Show that the mean matrix is $\mathbf{M} = (m_{ij}) = (p_{ij}\xi_j)$, where ξ_i is the expected number of offspring from a type i individual. Deduce that $\mathbf{M}$ is positive regular.

(c) If $\xi_i \equiv \xi$, so there is no natural selection in the population, show that $\rho = \xi$ and that the corresponding eigenvector (the stationary distribution) is the Hardy–Weinberg law $(p^2, 2pq, q^2)^T$.

11. Let (Z_n) be a supercritical BGW process, and denote by $Z_n^{(1)}$ the number of individuals in Z_n with infinite line of descent, and by $Z_n^{(2)}$ the number with finite line of descent.

(a) Show that $Z_n = (Z_n^{(1)}, Z_n^{(2)})$ is a two-type BGW process with

$$\mathbf{M} = \begin{pmatrix} m & (m-\nu)q/(1-q) \\ 0 & \nu \end{pmatrix}$$

where $\nu = P'(q) < 1$.

(b) Deduce that the process

$$Z_n(A) = \{Z_n(\omega)1(\omega) \in A)\}$$

defined on $(A, \mathcal{F}_A, \mathbf{P}_A)$, where $\mathbf{P}_A(F) = \mathbf{P}(AF)/\mathbf{P}(A)$, and A is the nonextinction set, is a supercritical BGW process.

(c) Using the result in (b), describe the asymptotic behavior of $\bar{m}_n$ on the nonextinction set.

APPENDIX

Some Data Sets

THE WHOOPING CRANE POPULATION OF NORTH AMERICA

Miller et al. (1974) report the annual counts from 1938 to 1972 of whooping cranes arriving at their wintering grounds in Aransas National Wildlife Refuge, Texas, USA. The young of the year have a distinctive plumage, and are therefore reported separately. In addition, we have inferred the number of deaths during the previous year, by subtracting the current year adults from the previous year total.

Year	Young	Adult	Total	Dead	Year	Young	Adult	Total	Dead
1938	4	10	14		1956	2	22	24	6
1939	6	16	22	(−2)	1957	4	22	26	2
1940	5	21	26	1	1958	9	23	32	3
1941	2	13	15	13	1959	2	31	33	1
1942	4	15	19	0	1960	6	30	36	3
1943	5	16	21	3	1961	5	33	38	3
1944	3	15	18	6	1962	0	32	32	6
1945	3	14	17	4	1963	7	26	33	6
1946	3	22	25	(−5)	1964	10	32	42	1
1947	6	25	31	0	1965	8	36	44	6
1948	3	27	30	4	1966	5	38	43	6
1949	4	30	34	0	1967	9	39	48	6
1950	5	26	31	8	1968	6	44	50	4
1951	5	20	25	9	1969	8	48	56	2
1952	2	19	21	6	1970	6	51	57	5
1953	3	21	24	0	1971	5	51	56	6
1954	0	21	21	3	1972	5	46	51	10
1955	8	20	28	1					

A SMALLPOX EPIDEMIC IN ABAKALIKI, NIGERIA

A total of 30 cases were observed in a population of 120 individuals at risk (Becker, 1976). The removal times, in days from the first removal, were

0							
13							
20	22	25	25	25	26	30	
35	38	40	40	42	42		
47	50	51					
55	55	56	57	58	60	60	61
66	66	71	76				

In this table we have divided the removal times in clusters about multiples of 12 days, the average length of the infectious period, yielding the (approximate) generation sizes

Generation	0	1	2	3	4	5	6	7
Size	1	1	7	6	3	8	4	0

NUMBER OF GOLD PARTICLES IN A FIXED VOLUME OF COLLOID SOLUTION

The following are the data from Westgren's (1916) experiment C, where he counted the number of gold particles in a small volume of water every 1.39 s (as determined by a metronome). The missing data at counts number 123, 238, 423, 479, 559, 580, 601, 657, 991, 1174, and 1410, have been estimated using the maximum of the conditional distribution given the previous observation.

2	1	1	1	1	1	0	2	2	1	1	1	2	3	2	3	0	0	0	0	0	0	1	1	0
0	1	0	1	1	1	2	2	3	3	4	5	3	4	2	2	1	2	1	3	2	0	2	2	1
0	2	2	2	1	2	3	2	2	2	3	2	2	2	2	2	2	1	3	3	4	2	2	1	1
0	1	2	0	0	0	0	1	1	1	2	1	1	0	1	2	1	1	1	2	1	2	2	1	2
1	1	1	2	1	1	1	2	1	1	1	0	1	1	1	0	0	1	0	0	1	1	1	1	1
1	1	1	0	2	0	1	1	0	1	0	1	1	0	0	1	2	2	2	1	2	1	2	1	2
2	3	0	1	0	0	0	2	2	2	2	2	1	1	0	1	2	1	3	4	5	4	6	5	4
0	1	1	1	2	1	1	1	0	2	0	0	1	1	1	1	2	2	2	1	1	2	3	2	3
4	3	3	4	2	4	3	3	3	2	1	2	1	3	2	1	1	0	1	2	1	0	0	0	0
0	1	0	0	0	2	3	4	2	2	2	3	2	2	3	0	1	1	1	3	5	2	2	2	3
3	2	2	2	1	2	3	4	3	0	0	1	1	1	1	1	2	1	1	1	2	1	1	1	2
3	2	4	3	4	0	3	0	5	4	4	5	3	4	6	5	2	4	5	4	2	4	1	2	3
2	1	2	4	3	2	3	3	4	5	7	5	4	6	3	2	5	4	3	5	4	3	2	3	3
3	4	3	3	2	1	2	2	3	2	3	4	3	2	2	1	1	3	3	3	3	3	3	3	2
2	2	0	1	1	1	0	1	1	1	0	0	0	0	0	1	2	0	1	1	2	1	0	3	1
2	2	1	1	1	1	0	1	1	1	0	0	0	1	0	0	0	0	0	0	1	0	1	1	1
1	2	3	2	1	1	1	3	5	4	2	4	3	2	2	2	2	3	3	2	3	1	1	3	2
2	2	2	1	5	3	4	4	3	3	2	1	3	3	2	0	1	3	3	2	1	2	5	3	4

5 3 3 4 3 2 3 2 3 0 0 1 1 2 2 1 2 3 2 3 1 2 2 3 2
3 2 1 1 3 2 0 2 1 3 3 3 5 4 2 1 3 4 3 3 4 4 5 5 3
1 0 2 4 4 4 5 3 3 2 1 0 2 1 2 2 3 2 1 0 1 0 2 2 1
1 1 2 3 3 1 1 2 1 1 2 1 0 2 2 1 2 1 0 1 1 1 0 0 1
2 1 0 1 2 2 3 2 2 3 4 2 1 1 2 1 2 1 1 0 2 1 1 0 1
1 1 1 0 0 3 1 3 2 1 1 1 1 1 1 1 1 1 0 0 0 0 0 0 0
0 2 1 2 2 2 1 2 0 0 1 2 1 1 2 1 1 0 0 2 1 1 1 0 0
0 0 0 0 0 0 1 1 1 1 0 0 1 1 1 1 1 1 2 1 1 0 1 2 0
0 0 1 1 1 0 0 1 1 1 1 1 1 1 1 1 2 2 1 2 2 1 1 1 2
2 1 0 1 0 0 0 0 2 1 0 0 1 2 1 2 1 0 1 1 0 1 2 2 1
3 0 0 3 2 3 2 2 4 2 3 1 2 1 0 2 3 2 2 3 1 3 4 5 4
2 3 2 2 2 1 1 1 3 2 3 3 2 4 4 4 3 3 2 3 0 2 0 2 1
2 3 5 6 5 7 5 6 4 2 1 3 1 0 2 4 2 3 1 3 1 2 3 2 0
1 1 1 3 2 5 3 3 2 1 3 2 4 1 4 5 4 2 4 3 2 4 3 0 1
0 2 3 2 4 3 4 2 0 0 2 3 1 2 3 2 2 1 0 1 0 0 2 2 2
0 2 0 3 1 0 3 2 2 2 3 2 0 1 0 0 0 0 0 0 0 0 1 1 1
0 1 0 1 1 1 0 0 2 1 2 1 2 3 2 1 1 2 3 3 3 3 2 3 2
1 2 2 1 2 1 1 0 0 0 0 0 0 1 1 0 1 1 1 1 1 0 0 1 0
0 1 1 1 1 0 0 0 0 0 0 1 0 0 0 1 0 1 1 1 1 0 1 1 1
0 1 0 1 0 1 1 0 0 0 0 0 0 0 0 0 0 1 1 1 0 0 0 0 1
1 1 0 0 0 0 0 0 1 0 1 0 1 2 1 1 2 1 1 0 2 2 1 2 1
1 1 1 1 0 0 1 1 0 0 0 0 1 1 1 1 0 1 0 2 2 1 1 1 1
0 0 0 0 0 0 2 1 0 0 2 3 2 2 3 2 3 2 2 3 2 1 1 1 0
0 0 0 0 0 0 0 0 0 0 2 2 1 2 1 1 0 2 2 1 1 0 1 1 1
2 0 1 1 1 1 0 0 0 1 1 2 1 1 0 0 0 0 0 0 0 1 0 0 1
2 1 2 1 1 3 1 1 1 1 1 1 0 0 1 1 1 2 2 1 2 3 2 1 1
2 1 1 0 0 1 1 1 0 1 3 2 1 1 1 2 3 2 2 0 1 0 0 0 1
0 1 1 0 0 0 0 0 1 1 1 0 0 2 1 1 1 1 1 1 1 1 2 1 1
1 1 0 1 1 0 1 2 2 2 1 0 0 1 1 3 3 4 4 3 2 1 1 1 1
1 1 0 0 0 1 0 0 0 0 0 0 0 0 0 0 0 0 1 2 1 1 2 0 1
0 1 0 0 3 4 3 2 0 1 2 2 2 1 2 1 2 2 1 1 1 1 1 1 0
1 1 1 1 1 1 0 0 0 1 1 0 0 0 0 0 0 0 0 0 0 0 1 1 0
0 0 1 2 1 0 1 1 1 1 0 0 0 0 0 1 2 0 1 1 1 1 0 1 0
1 1 1 2 0 0 0 0 2 4 4 4 3 1 1 1 1 1 1 1 1 1 1 1 0
1 1 1 1 0 1 1 0 1 2 2 1 1 0 1 1 1 2 1 2 2 1 0 1 1
0 0 2 1 2 0 1 1 1 0 1 3 3 2 3 3 3 4 3 4 4 3 5 4 4
3 2 3 3 3 1 2 2 2 1 0 1 2 2 2 1 1 1 1 2 3 0 0 0 0
0 1 1 2 2 2 1 2 1 2 0 0 1 2 2 1 1 1 1 2 1 1 1 2 2
2 2 2 2 1 2 2 1 2 1 1 1 0 3 1 1 1 1 0 1 0 1 1 3 1
2 1 2 2 1 1 2 1 2 2 1 2 2 2 2 0 1 2 0 0 0 1 1 2 2
3 2 3 2 1 1 0 1 1 1 1 0 1 3 2 3 2 3 4 5 3 3 2 0 1
2 3 2 1 3 2 3 2 2 1 1 0 2 1 1 2 1 3 2 2 2 2 2 1 1
2 3 2 2 1 0 1 1 1 1 0 0 1 3 2 2 2 2 1 2 2 1 1 1 0
2 1 2 1 3 2 1 1 0 0 1 1 1 0 0 0 0 0 0 1 0 0 0 0 0
0 0 0 0 0 0 0 1 1 2 2 2 1 1 1 1 1 1 1 1 0 1 2 1 0
0 0 0 0 0 0 0 0 1 1 0 1 0 0 1 2 2 1 1 1 0 1 1

SMALLPOX EPIDEMICS IN EUROPE

The number of smallpox importations into Europe during January 1961 to September 1970, resulting in k cases, as given by the World Health Organization.

k	1	2	3	4	5	>5
Observed number	12	3	4	1	1	9

PROGESTERONE CONCENTRATION IN THE BLOOD OF A PREGNANT WOMAN

The progesterone concentration in the blood was measured weakly for a healthy pregnant subject. These data were obtained from Fig. 2 in Winkel et al. (1976).

Week	Concentration (μg/I plasma)	Week	Concentration (μg/I plasma)	Week	Concentration (μg/I plasma)
12	11.79	19	16.19	26	36.84
13	12.95	20	20.00	27	36.84
14	12.80	21	24.42	28	39.07
15	13.41	22	26.83	29	40.00
16	13.26	23	33.15	30	44.46
17	14.74	24	33.54	31	46.06
18	16.97	25	35.56		

DISTRIBUTION OF ESTIMATED COMPLETED FRATERNITIES IN THE UNITED STATES IN 1928

The following are data given in Burks (1933) based on US birth statistics, excluding five states that did not report birth orders, and 10,000 cases for which birth order was unknown. The second column records the number N_k of individuals with birth order k born in 1928. There are a total of 2,086,882 births recorded.

Size	N_k	Size	N_k	Size	N_k
1	653820	10	27144	19	139
2	452383	11	16947	20	78
3	303510	12	10635	21	26
4	207452	13	6042	22	18
5	144793	14	3417	23	18
6	104447	15	1755	24	11
7	76218	16	907	25	6
8	55667	17	469	26	3
9	38706	18	271		

THE DARIUSLEUT AND LEHRERLEUT HUTTERITE POPULATIONS

The Hutterite Brethren comprise a Caucasian, religious and genetic isolate (Morgan and Holmes, 1982). They live on communal farms, called colonies, in the four western Canadian provinces and in the states of Washington, Montana, Minnesota, North Dakota, and South Dakota of the United States.

The origin of the Hutterite religion was the Anabaptist movement in the Tyrol region of Austria in the early 1500s. The majority of the Hutterite Brethren migrated from the Ukraine to North America during 1874–1879. The population is essentially closed to in-migration, and practically all marriages are within the group.

The offspring data here correspond to all 1236 married Dariusleut or Lehrerleut Hutterite women born between 1879 and 1936 with at least one child. The columns correspond to daughters, and the rows to sons.

	0	1	2	3	4	5	6	7	8	9	10	11	12
0		28	8	2	3	2	2	0	0	0	0	1	0
1	21	29	21	11	5	7	6	4	1	0	0	0	1
2	11	27	22	21	21	14	10	15	5	3	1	0	0
3	6	16	27	20	35	29	18	12	10	2	2	0	0
4	9	10	20	21	39	28	30	24	10	2	5	1	0
5	3	7	22	22	40	17	18	23	16	7	2	0	0
6	2	9	15	16	27	26	26	17	10	4	1	0	0
7	1	4	7	27	19	20	16	7	2	2	0	0	0
8	0	3	12	14	12	7	10	5	3	1	0	0	0
9	0	2	4	8	11	4	5	2	0	0	0	0	0
10	0	1	1	2	3	2	2	1	1	0	0	0	0
11	0	0	1	2	0	1	0	0	0	0	0	0	0
12	0	0	1	1	0	0	0	0	0	0	0	0	0

References

Adès, M., J.-P. Dion, G. Labelle, and K. Nanthi (1982): Recurrence formula and the maximum likelihood estimation of the age in a simple branching process. *J. Appl. Prob.* **19**: 776–784.

Albertsen, K. (1976): Slægternes uddøen. *Nord. Mat. Tidsskr.* **24**: 5–13.

Albertsen, K., and E. Kristensen (1976): En hidtil utrykt løsning på Erlang's problem om slægternes uddøen af Carl Marius Christensen. *Nord. Mat. Tidsskr.* **24**: 14–18.

Asmussen, S. (1976): Almost sure behavior of linear functionals of supercritical branching processes. *Trans. Amer. Math. Soc.* **231**: 233–248.

Asmussen, S. (1982): On the role of a certain eigenvalue in estimating the growth rate of a branching process. *Austral. J. Statist.* **24**: 151–159.

Asmussen, S., and H. Hering (1983): *Branching Processes*. Boston: Birkhäuser.

Asmussen, S., and N. Keiding (1978): Martingale central limit theorems and asymptotic estimation theory for multitype branching processes. *Adv. Appl. Prob.* **10**: 109–129.

Athreya, K. B. (1971): Some refinements in the theory of supercritical multitype branching processes. *Z. Warscheinlichkeitstheor. Verw. Geb.* **20**: 47–57.

Athreya, K. B., and S. Karlin (1971): On branching processes with random environments I. *Ann. Math. Statist.* **42**: 1499–1520.

Athreya, K. B., and P. E. Ney (1972): *Branching Processes*. Berlin: Springer.

Bailey, N. T. J. (1975): *The Mathematical Theory of Infectious Diseases and Its Applications*, 2nd Ed. London: Griffin.

Barndorff–Nielsen, O. (1978): *Information and Exponential Families*. New York: Wiley.

Bartholomew, D. J. (1967): *Stochastic Models for Social Processes*. London: Wiley.

Bartlett, M. S. (1966): *An Introduction to Stochastic Processes* (2nd Ed.). Cambridge: Cambridge University Press.

Bartoszyński, R. (1967): Branching processes and the theory of epidemics. In *Proc. Fifth Berkeley Symp. Math. Stat. Prob.*, eds. L. Le Cam and J. Neyman, **IV**: 259–269. University of California Press: Berkeley.

Basawa, I. V. (1981): Efficient conditional test for mixture experiments with applications to the birth and branching processes. *Biometrika* **68**: 153–164.

Basawa, I. V. (1987): Asymptotic distributions for prediction errors and related tests of fit for nonstationary processes. *Ann. Statist.* **15**: 46–58.

Basawa, I. V., and D. J. Scott (1976): Efficient tests for branching processes. *Biometrika* **63**: 531–536.

Basawa, I. V., and D. J. Scott (1983): *Asymptotic Optimal Inference for Non-Ergodic Models*. New York: Springer.

Bay, Z. (1941): Electron multiplier as an electron counting device. *Rev. Sci. Instr.* **12**: 127–133.

Becker, N. (1972): Vaccination programmes for rare infectious diseases. *Biometrika* **59**: 443–453.

Becker, N. (1974): On parametric estimation for mortal branching processes. *Biometrika* **61**: 393–399.

Becker, N. (1976): Estimation for an epidemic model. *Biometrics* **32**: 769–777.

Becker, N. (1977): Estimation for discrete-time branching processes with applications to epidemics. *Biometrics* **33**: 515–522.

Bharucha–Reid, A. T. (1956): On the stochastic theory of epidemics. In *Proc. Third Berk. Symp. Prob. Statist.*, ed. J. Neyman, **4**, 111–119.

Bhat, B. R., and S. R. Adke (1981): Maximum likelihood estimation for branching processes with immigation. *Adv. Appl. Prob.* **13**: 498–509.

Bienaymé, I. J. (1845): De la loi de multiplication et de la durée des familles. *Soc. Philomath. Paris* **5**: 37–39.

Billingsley, P. (1962): *Statistical Inference for Markov Chains*. Chicago: University of Chicago Press.

Billingsley, P. (1968): *Weak convergence of Probability Measures*. New York: Wiley.

Blake, J. (1989): Number of siblings and educational attainment. *Science* **245**: 32–36.

Borel, E. (1942): Sur l'emploi du théorème de Bernoulli pour faciliter le calcul d'un infinité de coefficients. Application au problème de l'attente à un guichet. *C. R. Acad. Sci. Paris* **214**: 452–456.

Burks, B. S. (1933): A statistical method for estimating the distribution of sizes of completed fraternities in a population represented by a random sampling of individuals. *J. Amer. Statist. Assoc.* **28**: 388–394.

Bühler, W. J. (1969): Ein zentraler Grenzwertsatz für Verzweigungsprozesse. *Z. Wahrscheinlichkeitstheor. Verw. Geb.* **11**: 139–141.

Chandrasekhar, S. (1954): Stochastic problems in physics and astronomy. In *Selected Papers on Noise and Stochastic Processes*, ed. N. Wax: 3–91. Dover: New York.

Châtauneuf, L. F. B. de (1847): Mémoire sur la durée des familles nobles de France. *Memoire de l'Académie royale des sciences morale et politiques de l'Institut de France* **5**: 753–794.

Cohn. H. (1982): Another look at the finite mean supercritical Bienaymé–Galton–Watson process. *J. Appl. Prob.* **19**: 307–312.

Cox. D. R. and D. V. Hinkley (1974): *Theoretical Statistics*. London: Chapman and Hall.

Cox. D. R., and H. D. Miller (1965): *The Theory of Stochastic Processes*. London: Chapman and Hall.

Crump, K. S., and J. H. Gillespie (1976): The dispersion of a neutral allele considered as a branching process. *J. Appl. Prob.* **13**: 208–218.

Crump, K. S., and R. B. Howe (1972): Nonparametric estimation of the age of a Galton–Watson branching process. *Biometrika* **59**: 533–538.

Cryer, J. D., and T. Robertson (1975): Isotonic estimation of the probability of extinction of a branching process. *J. Amer. Statist. Assoc.* **70**: 905–912.

Daly, F. (1979): Collapsing supercritical branching processes. *J. Appl. Prob.* **16**: 732–739.

Davies, R. B. (1985): Asymptotic inference when the amount of information is random. In *Proc. Berk. Conf. in Honor of Jerzy Neyman and Jack Kiefer*, **II**: 841–864. Monterey: Wadsworth.

Dempster, A. P., N. M. Laird, and D. B. Rubin (1977): Maximum likelihood from incomplete data via the EM algorithm. *J. Roy. Statist. Soc. Ser. B* **39**: 1–38.

de Candolle, A. (1873): *Histoire des Sciences et des Savants Depuis Deux Siècles*. Genéve: Georg.

Diaconis, P., and D. Ylvisaker (1979): Conjugate priors for exponential families. *Ann. Statist.* **7**: 269–281.

Dion, J.-P. (1972): Estimation des probabilités initiales et de la moyenne d'un processus de Galton–Watson. Thèse de Doctorat, Université de Montréal, Montréal.

Dion, J.-P. (1974). Estimation of the mean and the initial probabilities of a branching process. *J. Appl. Prob.* **11**: 687–694.

Dion. J.-P. (1975): Estimation of the variance of a branching process. *Ann. Statist.* **3**: 1183–1187.

Dion, J.-P., and N. Keiding (1978): Statistical inference in branching processes. In *Branching Processes*, eds. A. Joffe and P. Ney: 105–140. New York: Dekker.

Dion. J.-P., G. Labelle, and A. Latour (1982): Small sample results for maximum-likelihood estimation in branching processes. *Can. J. Statist.* **10**: 271–276.

Dion, J.-P., and W. W. Esty (1979): Estimation problems in branching processes with random environments. *Ann. Statist.* **7**: 680–685.

Dobzhansky, T., and S. Wright (1947): Genetics of natural populations. XV. Rate of diffusion of a mutant gene through a population of Drosophila pseudoobscura. *Genetics* **32**: 303–324.

Dubuc, S. and E. Seneta (1976): The local limit theorem for the Galton–Watson process. *Ann. Prob.* **4**: 490–496.

Duby, C., and A. Rouault (1982): Estimation non paramétrique de l'espérance et de la variance de la loi de reproduction d'un processus de ramification. *Ann. Inst. Henri Poincarè* **XVIII B**: 149–163.

Durbin, J. (1960): Estimation of parameters in time-series regression models. *J. Roy. Statist. Soc.* **B22**: 139.

Dvoretzky, A., and J. Wolfowitz (1951): Sums of random integers reduced modulo m. *Duke Math. J.* **18**: 501–507.

Efron, B. (1975): Defining the curvature of a statistical problem (with applications to second order efficiency). *Ann. Statist.* **3**: 1189–1242.

Efron, B., and D. V. Hinkley (1978): Assessing the accuracy of the maximum likelihood estimator: Observed versus expected Fisher information. *Biometrika* **65**: 457–487.

Einstein, A. (1905): Über die von der molekular-kinetischen Theorie der Wärme geforderte Bewegung von in ruhenden Flüssigkeiten suspendierten Teilchen. *Ann. Phys.* **17**: 549–560.

Einstein, A. (1956): *Investigation on the Theory of the Brownian Movement.* New York: Dover.

Engle, R. F. (1982): Autoregressive conditional heteroscedasticity with estimates of the variance of United Kingdom inflation. *Econometrica* **50**: 987–1007.

Erlang, A. K. (1929): Opgave Nr. 15. *Mat. Tidsskr. B*: 36.

Eschenbach, W., and W. Winkler (1975): Maximum-Likelihood-Schatzungen beim Verzweigungsprozeß von Galton–Watson. *Math. Op. Forsch. Statist.* **6**: 213–224.

Fahlbeck, P. E., *Sveriges Adel 1*, Lund, 1898.

Feigin, P. D. (1977): A note on maximum likelihood estimation for simple branching processes. *Austr. J. Statist.* **19**: 152–154.

Feigin, P. D. (1978): The efficiency criteria problem for stochastic processes. *Stoch. Proc. Appl.* **6**: 115-128.

Feigin, P. D. (1986): Asymptotic theory of conditional inference for stochastic processes. *Stoch. Proc. Appl.* **22**: 89–102.

Feigin, P. D., and B. Reiser (1979): On asymptotic ancillarity and inference for Yule and regular nonergodic processes. *Biometrika* **66**: 279–283.

Feller, W., (1968): *An Introduction to Probability Theory and its Applications* (3rd Ed.), **I**. New York: Wiley.

Ferguson, T. S. (1973): A Bayesian analysis of some nonparametric problems. *Ann. Statist.* **1**: 209–230.

Finetti, B. de (1931): Funzione caratteristica di un fenomeno aleatorio. *Atti R. Acad. Naz. Lincei* **4**: 251–299.

Fisher, R. A. (1922a): On the dominance ratio. *Proc. R. Soc. Edinburgh* **42**: 321–341.

Fisher, R. A. (1922b): On the mathematical foundations of theoretical statistics. *Philos. Trans. A* **222**: 309–368.

Fisher, R. A. (1930): *The Genetical Theory of Natural Selection.* Oxford: Oxford University Press.

Fisher, R. A. (1958): *The Genetical Theory of Natural Selection* (Rev. Ed.). New York: Dover.

Fisher, R. A. (1973): *Statistical Methods and Scientific Inference.* New York: Hafner Press.

Fraser, D. A. S., and P. McDunnough (1984): Further remarks on asymptotic normality of likelihood and conditional analyses. *Can. J. Statist.* **12**: 183–190.

Freedman, D. (1971): *Markov Chains.* San Francisco: Holden–Day.

Galton, F. (1869): *Hereditary Genius*, London.

Galton, F. (1873): Problem 4001. *Educ. Times* **1**: 17.

Gani, J., and I. W. Saunders (1976): On the parity of individuals in a branching process. *J. Appl. Prob.* **13**: 219–230.

Gani, J. and I. W. Saunders (1977): Fitting a model to the growth of yeast colonies. *Biometrics* **33**: 113–120.

Genet, R. M., D. S. Hayes, D. S. Hall, and D. R. Genet (1987): *Supernova 1987A—Astronomy's Explosive Enigma*. Fairfield: Fairborn Press.

Gilberg, A., L. Gilberg, R. Gilberg, and M. Holm. (1978): Polar Eskimo genealogy. In *Meddelelser om Grønland* **203**. København: Arnold Busck.

Godambe, V. P. (1960): An optimum property of regular maximum likelihood estimation. *Ann. Math. Statist.* **31**: 1208–1211.

Godambe, V. P. (1985): The foundations of finite sample estimation in stochastic processes. *Biometrika* **72**: 419–428.

Godambe, V. P., and C. C. Heyde (1987): Quasi-likelihood and optimal estimation. *Inst. Statist. Rev.* **55**: 231–244.

Good, I. J. (1949): The number of individuals in a cascade process. *Proc. Cambridge Philos. Soc.* **45**: 360–363.

Good, I. J. (1957): Distribution of word frequencies. *Nature* **179**: 595.

Good, I. J. (1963): Cascade theory and the molecular weight average of the sol fraction. *Proc. Roy. Soc. London* **A 272**: 54–59.

Gordon, M. (1962): Good's theory of cascade processes applied to the statistics of polymer distributions. *Proc. Roy. Soc. London* **A 268**: 240–259.

Guttorp, P., and R. A. Lockhart (1989): Estimation in sparsely sampled random walks. *Stoch. Proc. Appl.* **31**: 315–320.

Guttorp, P., and A. F. Siegel (1985): Consistent estimation in partially observed random walks. *Ann. Statist.* **13**: 958–969.

Haldane, J. B. S. (1927): A mathematical theory of natural and artificial selection. Part V: Selection and mutation. *Proc. Cambridge Philos. Soc.* **23**: 838–844.

Hall, P. (1982): *Rates of Convergence in the Central Limit Theorem*. Boston: Pitman.

Hall, P., and C. C. Heyde (1980): *Martingale Limit Theory and its Applications*. New York: Academic Press.

Hannum, R. C., M. Hollander, and N. Langberg (1981): Distributional results for random functionals of a Dirichlet process. *Ann. Prob.* **9**: 665–670.

Harrington, D. P., and T. R. Fleming (1978): Estimation for branching processes with varying and random environments. *Math. Biosci.* **39**: 255–271.

Harris, T. E. (1948): Branching processes. *Ann. Math. Statist.* **19**: 474–494.

Harris, T. E. (1963): *Branching Processes*. New York: Springer.

Hájek, J. (1967): On basic concepts of statistics. In *Proc. Fifth Berkeley Symp. Math. Stat. Prob.*, eds. L. Le Cam and J. Neyman, **I**: 139–162. University of California Press: Berkeley.

Heyde, C. C. (1970): Extension of a result of Seneta for the super-critical Galton–Watson process. *Ann. Math. Statist.* **41**: 739–742.

Heyde, C. C. (1974): On estimating the variance of the offspring distribution in a simple branching process. *Adv. Appl. Prob.* **3**: 421–433.

Heyde, C. C. (1975): Remarks on efficiency in estimation for branching processes. *Biometrika* **62**: 49–55.

Heyde, C. C. (1978): On an optimal property of the maximum likelihood estimator of a parameter from a stochastic process. *Stoch. Proc. Appl.* **6**: 1–9.

Heyde, C. C. (1979): On assessing the potential severity of an outbreak of a rare infectious disease: A Bayesian approach. *Austral. J. Statist.* **21**: 282–292.

Heyde, C. C., and P. D. Feigin (1975): *On Efficiency and Exponential Families in Stochastic Process Estimation. Statistical Distributions in Scientific Work*, **I**: 227–240. Dordrecht: Reidel.

Heyde, C. C., and I. M. Johnstone (1979): On asymptotic posterior normality for stochastic processes. *J. Roy. Statist. Soc.* **B 41**: 184–189.

Heyde, C. C., and J. R. Leslie (1971): Improved classical limit analogues for Galton–Watson processes with or without immigration. *Bull. Austral. Math. Soc.* **5**: 145–156.

Heyde, C. C., and E. Seneta (1972a): Estimation theory for growth and immigation rates in a multiplicative process. *J. Appl. Prob.* **9**: 235–256.

Heyde, C. C. and E. Seneta (1972b): The simple branching process, a turning point test and a fundamental identity: a historical note on I. J. Bienaymé. *Biometrika* **59**: 680–683.

Heyde, C. C., and E. Seneta (1977): *I. J. Bienaymé—Statistical Theory Anticipated. Studies in the History of Mathematics and Physical Sciences*, **3**. New York: Springer.

Hudson, I. L. (1982): Large sample inference for Markovian exponential families with application to branching processes with immigration. *Austral. J. Statist.* **24**: 98–112.

Hwang, T.-Y., and J.-T. Hwang (1978): Maximum likelihood estimate of the age of a Galton–Watson process with Poisson offspring distribution. *Bull. Inst. Math. Acad. Sinica* **6**: 203–213.

Jagers, P. (1975): *Branching Processes with Biological Applications*. London: Wiley.

Jeffreys, H. (1983): *Theory of Probability* (3rd Ed.). Oxford: Clarendon Press.

Jeganathan, P. (1982): On the asymptotic theory of estimation when the limit of Log–Likelihood ratios is mixed normal. *Sankhya* **A44**: 173–212.

Joffe, A., and W. A. O'N. Waugh (1982): Exact distribution of kin numbers in a Galton–Watson process. *J. Appl. Prob.* **19**: 767–775.

Johnson, R. A., V. Susarla, and J. van Ryzin (1979): Bayesian non-parametric estimation for age-dependent branching processes. *Stoch. Proc. Appl.* **9**: 307–318.

Kallenberg, P. J. M. (1979): *Branching Processes with Continuous State Space*, Mathematical Centre Tracts: 117. Amsterdam.

Karlin, S., and J. McGregor (1967): Properties of the stationary measure of the critical case simple branching process. *Ann. Math. Statist.* **38**: 977–991.

Kedem, B., and L. S. Chiu (1987): On the lognormality of rain rate. *Proc. Natl. Acad. Sci.* **84**: 901–905.

Keiding, N. (1974): Estimation in the birth process. *Biometrika* **7**: 71–80, 647.

Keiding, N. (1975a): Estimation theory for branching processes. *Bull. Inst. Statist. Inst.* **46**: 12–19.

Keiding, N. (1975b): Extinction and exponential growth in random environments. *Theor. Pop. Biol.* **8**: 49–63.

Keiding, N. (1976): Population growth and branching processes in random environments. In *Proc. 9th Int. Biometric Conf.*: 149–165. Boston.

Keiding, N., and S. Lauritzen (1978): Marginal maximum likelihood estimates and estimation of the offspring mean in a branching process. *Scand. J. Statist.* **5**: 106–110.

Keiding, N., and J. E. Nielsen (1973): The growth of supercritical branching processes with random environments. *Ann. Prob.* **1**: 1065–1067.

Kendall, D. G. (1951): Some problems in the theory of queues. *J. Roy. Statist. Soc.* **B 13**: 151–185.

Kendall, D. G. (1966): Branching processes since 1873. *J. London Math. Soc.* **41**: 385–406.

Kendall, D. G. (1975): The genealogy of genealogy. Branching processes before (and after) 1873. *Bull. London Math. Soc.* **7**: 225–253.

Kesten, H., P. Ney, and F. Spitzer (1966): The Galton–Watson process with mean one and finite variance. *Theory Prob. Appl.* **11**: 513–540.

Kesten, H., and B. P. Stigum (1966): Additional limit theorems for indecomposable multidimensional Galton–Watson processes. *Ann. Math. Statist.* **37**: 1463–1481.

Keyfitz, N. (1968): *Introduction to the Mathematics of Population*. Reading.

Keyfitz, N. (1985): *Applied Mathematical Demography* (2nd Ed.). New York: Springer.

Keyfitz, N., and A. Tyree (1967): Computerization of the branching process. *Behav. Sci.* **12**: 329–336.

Klimko, L. A., and P. I. Nelson (1978): On conditional least squares estimation for stochastic processes. *Ann. Statist.* **6**: 629–642.

Kojima, K.-I., and T. M. Kelleher (1962): Survival of mutant genes. *Amer. Naturalist* **XCVI**: 329–343.

Kurrelmeyer, B., and L. J. Hayner (1937): Shot effect of secondary electrons from nickel and beryllium. *Phys. Rev.* **52**: 952–958.

Latour, A. (1981): Le nombre de réalisations possibles d'un processus de Galton–Watson. *Ann. Sci. Math. Québec* **5**: 137–145.

Lauritzen, S. L. (1982): *Statistical Models as Extremal Families*. Aalborg: Aalborg University Press.

Lauritzen, S. L. (1984): Extreme point models in statistics. *Scand. J. Statist.* **11**: 65–91.

Lauritzen, S. (1988): *Extremal Families and Systems of Sufficient Statistics*. New York: Springer.

Lehmann, E. L. (1959): *Testing Statistical Hypotheses*. New York: Wiley.

Le Cam, L. (1986): *Asymptotic Methods in Statistical Decision Theory*. New York: Springer.

Lipow, C. (1971): Two branching models with generating functions dependent on population size. Ph.D. dissertation, University of Wisconsin.

Lockhart, R. A. (1982): On the nonexistence of consistent estimates in Galton–Watson branching processes. *J. Appl. Prob.* **19**: 842–846.

Lootgieter, J. C. (1977): La σ-algébre asymptotique d'une chaîne de Galton–Watson. *Ann. Inst. H. Poincaré* **XIII**: 193–230.

Lotka, A. J. (1931): The extinction of families, I. *J. Wash. Acad. Sci.* **21**: 377–380.

Lotka, A. J. (1939): Theorie analytiques des associations biologiques. *Act. Sci. Indust.* **2**: 123–136.

Maki, E., and P. McDunnough (1989a): Sampling a branching tree. *Stoch. Proc. Appl.* **31**: 283–305.

Maki, E. and P. McDunnough (1989b): What can and can't be estimated in branching and related processes? *Stoch. Proc. Appl.* **31**: 307–314.

Malthus, T. R. (1817): *An Essay on the Principle of Population, as it Affects the Future Improvements of Society, with Remarks on the Speculations of Mr. Godwin, M. Condorcet, and Other Writers* (5th Ed.). London: John Murray.

McKendrick, A. G. (1926): Application of mathematics to medical problems. *Proc. Edinburgh Math. Soc.* **44**: 98–130.

Meilijson, I. (1989): A fast improvement of the EM algorithm on its own terms. *J. R. Statist. Soc. B* **51**: 127–138.

Miller, R. S., D. B. Botkin, and R. Mendelssohn (1974): The whooping crane (*Grus americana*) population of North America. *Biol. Cons.* **6**: 106–111.

Moran, P. A. P. (1962): *The Statistical Processes of Evolutionary Theory*. Oxford: Clarendon Press.

Morgan, K., and T. M. Holmes (1982): Population Structure of a religious isolate: The Dariusleut Hutterites of Alberta. In *Current Developments in Anthropological Genetics*, eds. Cruawford and Mielke, New York: Plenum Press.

Nanthi, K. (1979): Estimation of the criticality parameter of a supercritical branching process with random environments. *J. Appl. Prob.* **16**: 890–896.

Nanthi, K. (1983): *Statistical Estimation for Stochastic Processes*, Queen's Papers in Pure and Applied Mathematics: 62.

Nedelman, J., J. A. Thompson, and R. J. Taylor (1987): The statistical demography of whooping cranes. *Ecology* **68**: 1401–1411.

Nemec, A. F. L. (1984): Conditionally heteroscedastic autoregressions. Technical Report 43, Department of Statistics, University of Washington, Seattle.

Otter, R. (1949): The multiplicative process. *Ann. Math. Statist.* **20**: 206–224.

Pakes, A. G. (1975): Non-parametric estimation in the Galton–Watson process. *Math. Biosciences* **26**: 1–18.

Pakes, A. G., and C. C. Heyde (1982): Optimal estimation of the criticality parameter of a supercritical branching process having random environments. *J. Appl. Prob.* **19**: 415–420.

Petrov, V. V. (1975): *Sums of Independent Variables*. Berlin: Springer.

Poincaré, H. (1890): Sur une classe nouvelle de trenscendantes uniformes. *J. Math. Pures Appl.* **6** (**4th Ser.**), 313.

Pollard, J. H. (1973): *Mathematical Models for the Growth of Human Populations*. Cambridge: Cambridge University Press.

Press, W. H., B. P. Flannery, S. A. Teukolsky, and W. T. Vetterling (1986): *Numerical Recipes: The Art of Scientific Computing*. Cambridge: Cambridge University Press.

Raiffa, H., and R. Schleiffer (1961): *Applied Statistical Decision Theory*. Cambridge, MA: MIT Press.

Rao, C. R. (1973): *Linear Statistical Inference and its Applications* (2nd Ed.). New York: Wiley.

Reid, N. (1988): Saddlepoint methods (with Discussion). *Statist. Sci.* **3**: 213–238.

Rida, W. N. (1991): Asymptotic properties of some estimators for the infection rate in the general stochastic epidemic model. *J. Roy. Statist. Soc. Ser B* **53**: 269–283.

Rodrigues da Silva, G., S. I. Rabello, and J. J. Angulo (1963): Epidemic of *Variola minor* in a suburb of São Paulo. *Public Health Report, Washington* **78**: 165–171.

Saunders, I. W. (1976): A convergence theorem for parities in a birth-and-death process. *J. Appl. Prob.* **13**: 231–238.

Schelling, H. von (1944): Das Alles oder Nichts–Gesetz, gedeutet als Endergebnis einer Auslösnungsfolge. *Abh. Preuss. Akad. Wiss., Math.-Naturw. Klasse* **6**.

Scott, D. (1987): On posterior asymptotic normality and asymptotic normality of estimators for the Galton–Watson process. *J. Roy. Statist. Soc.* **B49**: 209–214.

Scott, D. J. (1978): A central limit theorem for martingales and application to branching processes. *Stoch. Proc. Appl.* **6**: 241–252.

Seneta, E. (1969): Functional equations and the Galton–Watson process. *Adv. Appl. Prob.* **1**: 1–42.

Sheehan, N. (1990): Genetic restoration of complex pedigrees. Ph.D. dissertation, Dept. of Statistics, University of Washington, Seattle.

Shockley, W., and J. R. Pierce (1938): A theory of noise for electron multipliers. *Proc. Inst. Radio Eng.* **26**: 321–332.

Smoluchowski, M. v. (1916): Drei Vorträge über Diffusion, Brownsche Bewegung und Koagulation von Kolloidteilchen. *Phys. Z.* **17**: 557, 585.

Steffensen, J. F. (1930): Om Sandssynligheden for at Afkommet uddør. *Mat. Tidsskr.*: 19–23.

Stigler, S. M. (1970): Estimating the age of a Galton–Watson process. *Biometrika* **57**: 505–512.

Stigler, S. M. (1971): The estimation of the probability of extinction and other parameters associated with branching processes. *Biometrika* **58**: 499–508.

Sweeting, T. J. (1978): On efficient tests for branching processes. *Biometrika* **65**: 123–128.

Sweeting, T. J. (1986): Asymptotic conditional inference for the offspring mean of a supercritical Galton–Watson process. *Ann. Statist.* **14**: 925–933.

Swensen, A. R. (1980): Asymptotic inference for a class of stochastic processes. Ph. D. dissertation, Dept. of Statistics, University of California, Berkeley.

Swensen, A. R. (1983): A note on asymptotic inference in a class of non-stationary processes. *Stoch. Proc. Appl.* **15**: 181–191.

Sykes, Z. M. (1969): Some stochastic versions of the matrix model for population dynamics. *J. Amer. Statist. Assoc.* **64**: 111–130.

Thompson, E. A. (1976): Estimation of age and rate of increase of rare variants. *Am. J. Hum. Genet.* **28**: 442–452.

Tierney, L., and J. B. Kadane (1986): Accurate approximation for posterior moments and marginal densities. *J. Amer. Statist. Assoc.* **81**: 82–86.

Tierney, L., R. E. Kass, and J. B. Kadane (1987): Approximation of expectations and variances using Laplace's method. Technical report, Department of Statistics, Carnegie Mellon University.

Venkataraman, K. N. (1982): A time series approach to the study of the simple subcritical Galton–Watson process with immigration. *Adv. Appl. Prob.* **14**: 1–20.

Watson, H. W. (1873): Solution to Problem 4001. *Educ. Times* **1**: 115–116.

Watson, H. W., and F. Galton (1874): On the probability of extinction of families. *J. Anthropol. Inst. Great Britain and Ireland* **4**: 138–144.

Waugh, W. A. O'N. (1958): Conditioned Markov processes. *Biometrika* **45**: 241–249.

Waugh, W. A. O'N. (1981): Application of the Galton–Watson process to the kin number problem. *Adv. Appl. Proc.* **13**: 631–649.

Wei, C. Z. and J. Winnicki (1990): Estimation of the means in the branching process with immigation. *Ann. Statist.* **18**: 1757–1773.

Weinberg, W. (1914): Die Abnahme der Knabenziffer bei in männlicher Linie aussterbenden und erhaltenen Geschlechten. *Arch. Rassen Gesellschaftbiol.* **11**: 46–95.

Weiss, L., and J. Wolfowitz (1974): *Maximum Probability Estimators and Related Topics. Lecture Notes in Mathematics*, **424**. Berlin: Springer.

Westgren, A. (1916): Die Veränderungsgeschwindigkeit der lokalen Teilchenkonzentration in kolloiden Systemen (Erste Mitteilung). *Ark. Mat. Astron. Fys.* **11**: No. 14, 1–25.

Winkel, P., P. Gaede, and J. Lyngbye (1976): Method for monitoring plasma progesterone concentrations in pregnancy. *Clin. Chem.* **22**: 422–428.

Winnicki, J. (1988): Estimation theory for the branching process with immigration. *Contemporary Mathematics* **80**: 301–322.

Wu, C. F. J. (1983): On the convergence properties of the EM algorithm. *Ann. Statist.* **11**: 95–103.

Yaglom, A. M. (1947): Nekotorye predelnye teoremy teorii vetv̌yaščihsya slučaynyh processov (Some limit theorems from the theory of branching stochastic processes). *Dokl. Akad. Nauk SSSR* **56**: 795–798.

Yanev, N. M. (1975): On the statistics of branching processes. *Theory Prob. Appl.* **20**: 612–622.

Zhang, Y. C. (1983): A limit theorem for the continuous state branching process. *Stoch. Proc. Appl.* **14**: 175–186.

Zworykin, V. K., G. A. Morton, and L. Malter (1936): The secondary emission multiplier—a new electronic device. *Proc. Inst. Radio Eng.* **24**: 351–375.

Author Index

Subject Index

Applied Prob

HOEL a[illegible] *Third Edition*

HOGG and KLUGMAN • Loss Distributions

HOLLANDER and WOLFE • Nonparametric Statistic[illegible]

HOSMER and LEMESHOW • Applied Logistic Regre[illegible]

IMAN and CONOVER • Modern Business Statistics

JACKSON • A User's Guide to Principle Components

JESSEN • Statistical Survey Techniques

JOHN • Statistical Methods in Engineering and Quality Assurance

JOHNSON • Multivariate Statistical Simulation

JOHNSON and KOTZ • Distributions in Statistics
Discrete Distributions
Continuous Univariate Distributions—1
Continuous Univariate Distributions—2
Continuous Multivariate Distributions

JUDGE, GRIFFITHS, HILL, LÜTKEPOHL, and LEE • The Theory and Practice of Econometrics, *Second Edition*

JUDGE, HILL, GRIFFITHS, LÜTKEPOHL, and LEE • Introduction to the Theory and Practice of Econometrics, *Second Edition*

KALBFLEISCH and PRENTICE • The Statistical Analysis of Failure Time Data

KASPRZYK, DUNCAN, KALTON, and SINGH • Panel Surveys

KAUFMAN and ROUSSEEUW • Finding Groups in Data: An Introduction to Cluster Analysis

KEENEY and RAIFFA • Decisions with Multiple Objectives

KISH • Statistical Design for Research

KISH • Survey Sampling

KUH, NEESE, and HOLLINGER • Structural Sensitivity in Econometric Models

LAWLESS • Statistical Models and Methods for Lifetime Data

LEAMER • Specification Searches: Ad Hoc Inference with Nonexperimental Data

LEBART, MORINEAU, and WARWICK • Multivariate Descriptive Statistical Analysis: Correspondence Analysis and Related Techniques for Large Matrices

LEVY and LEMESHOW • Sampling of Populations: Methods and Applications

LINHART and ZUCCHINI • Model Selection

LITTLE and RUBIN • Statistical Analysis with Missing Data

McNEIL • Interactive Data Analysis

MAGNUS and NEUDECKER • Matrix Differential Calculus with Applications in Statistics and Econometrics

MAINDONALD • Statistical Computation

MALLOWS • Design, Data, and Analysis by Some Friends of Cuthbert Daniel

MANN, SCHAFER, and SINGPURWALLA • Methods for Statistical Analysis of Reliability and Life Data

MASON, GUNST, and HESS • Statistical Design and Analysis of Experiments with Applications to Engineering and Science

MILLER • Survival Analysis

MILLER, EFRON, BROWN, and MOSES • Biostatistics Casebook

MONTGOMERY and PECK • Introduction to Linear Regression Analysis, *Second Edition*

NELSON • Accelerated Testing, Statistical Models, Test Plans, and Data Analyses

NELSON • Applied Life Data Analysis

OCHI • Applied Probability and Stochastic Processes in Engineering and Physical Sciences

OSBORNE • Finite Algorithms in Optimization and Data Analysis

OTNES and ENOCHSON • Digital Time Series Analysis

PANKRATZ • Forecasting with Univariate Box-Jenkins Models: Concepts and Cases

PANKRATZ • Forecasting with Dynamic Regression Modes

POLLOCK • The Algebra of Econometrics

RAO and MITRA • Generalized Inverse of Matrices and Its Applications

RÉNYI • A Diary on Information Theory

RIPLEY • Spatial Statistics

RIPLEY • Stochastic Simulation

ROSS • Introduction to Probability and Statistics for Engineers and Scientists

ROUSSEEUW and LEROY • Robust Regression and Outlier Detection

RUBIN • Multiple Imputation for Nonresponse in Surveys

*Now available in a lower priced paperback edition in the Wiley Classics Library.